**WORTHEN**   **BORG**

# Student Exercise Manual

to accompany
Measurement and Evaluation
in the Schools

**Prepared by**

**Kenneth W. Merrell**
*Utah State University*

Longman

**Student Exercise Manual to accompany
Measurement and Evaluation in the Schools**

Copyright ©1993 by Longman

Longman, 10 Bank Street, White Plains, N.Y. 10606

Associated Companies:
Longman Group Ltd., London
Longman Cheshire Pty., Melbourne
Longman Paul Pty., Auckland
Copp Clark Pitman, Toronto

**ISBN 0-8013-1050-4**

2 3 4 5 6 7 8 9 10-CRS-99 98 97 96 95 94

# CONTENTS

| | | |
|---|---|---|
| To the Student | | 4 |

| | | |
|---|---|---|
| 1 | What Is Measurement and Why Study It? | 5 |
| 2 | How Did We Get Where We Are in Measurement? | 13 |
| 3 | Coming to Grips with Current Social, Legal, and Ethical Issues in Measurement | 21 |
| 4 | Getting Your Bearings: Some Concepts and Classifications Basic to Measurement and Evaluation | 29 |
| 5 | Learning to Read the Sign Posts: What Do Those Test Scores Really Mean? | 38 |
| 6 | Why Worry about Reliability? Reliable Measures Yield Trustworthy Scores | 46 |
| 7 | Why Worry about Validity? Valid Measures Permit Accurate Conclusions | 54 |
| 8 | Cutting Down Test Score Pollution: The Influence of Extraneous Factors | 62 |
| 9 | Constructing Your Own Achievement Tests--Deciding When and How to Do So | 70 |
| 10 | Steps in Developing Good Test Items for Your Achievement Tests | 78 |
| 11 | The Process of Becoming an Expert Tester: Assembly, Administration, and Analysis | 86 |
| 12 | Constructing and Using Descriptive Measures: Questionnaires, Interviews, Observations, and Rating Scales | 94 |
| 13 | Getting in Touch with Students' Feelings: Measuring Attitudes and Interests | 104 |
| 14 | Picking the Right Yardstick: Assigning Grades and Reporting Student Performance | 111 |
| 15 | Avoiding Being Caught in the Crossfire between Standardized Test Supporters and Alternative Assessment Enthusiasts | 118 |
| 16 | Finding and Selecting Measures That Can Help Solve Your Educational Problems | 126 |
| 17 | What Have My Students Learned? An Introduction to Standardized Achievement Measures | 134 |
| 18 | Keeping Your Finger on the Student's Pulse: Using Tests to Diagnose Strengths and Weaknesses | 142 |
| 19 | Assessing Your Students' Potential: A Look at Aptitude and Readiness Measures | 150 |
| 20 | Being Sensitive to Your Students' Personal Problems: Measures of Personal and Social Adjustment | 158 |
| 21 | Setting Up a School Testing Program | 167 |
| 22 | Setting Up a School Evaluation Program | 173 |

| | | |
|---|---|---|
| Notes | | 183 |

## TO THE STUDENT

This workbook is designed to help you understand and apply the concepts explained in *Measurement and Evaluation in the Schools: A Practical Guide.* By thoughtfully using this workbook as a supplement to your study, you will be able to maximize your knowledge and skills in measurement and evaluation.

Each chapter in the workbook begins with a short summary of information of the related text chapter, along with a listing of the major topical headings in that chapter. Three sets of *Guided Study Exercises* designed to help you master the essential concepts and terms of each chapter are included next. Answers to each of these exercises are located at the end of the chapter. The three sets of exercises are as follows:

*1. Definition Exercises.* These exercises include 10-15 key terms from the chapter for which you will need to provide a brief definition. The best way to do these exercises is to go back through the chapter and locate each term. Some of the terms will be easy to define, in that the textbook chapter defines them as they are introduced. Other terms will be more difficult to define and will require you to "read between the lines" by compiling information related to that term from various places in the chapter.

*2. Multiple-Choice Questions.* Eight multiple-choice questions are included for each chapter. These are designed to test your knowledge of the information in the text chapter in a more sophisticated way than the definition exercises. You will need to sort between the essential and nonessential details of certain concepts, be able to recognize practical applications of different measurement and evaluation techniques, and be able to match sentence stems to appropriate completions for those sentences.

*3. Short-Answer Questions.* The third set of guided study exercises includes five brief questions that will help you integrate some of the major concepts in the text chapter in more detail and at a higher cognitive level.

Each chapter also contains an *Application Activities* section to help you integrate and demonstrate your understanding of some of the key concepts in the text. In each of these exercises, you are asked to complete a practical task related to measurement and evaluation. These tasks portray realistic responsibilities a classroom teacher might be asked to assume. As with the three *Guided Study Exercises*, answers to these activities are found at the end of the workbook chapter.

At the end of the workbook, you will find a few blank pages for use in making notes and working out solutions to some of the exercises and activities. You may choose not to complete all of the different workbook activities, but reviewing the content and completing at least a few activities in each section of the workbook chapter will help you master the important concepts from the text.

# CHAPTER 1
# WHAT IS MEASUREMENT AND WHY STUDY IT?

## CHAPTER SUMMARY

As a college student, you undoubtedly have substantial experience taking educational tests. As an educator, measurement will likely be an important part of your work. This chapter introduces the concept of educational measurement and provides an overview of the different ways measurement is used in education, as well as some of the ways it is abused.

The chapter begins with a discussion of the role of measurement and evaluation in education. Specifically, the need for measurement and evaluation is discussed, and the concepts of subjective judgment, objective testing, and quantification are introduced. The majority of the chapter examines the various uses of educational measurement, including instructional decisions, entry-exit decisions, administrative and policy decisions, and expansion of knowledge.

This chapter will provide you with a foundation and focus for learning the specific technical material on measurement and evaluation in the remainder of the book. Study it carefully and become familiar with the issues raised. Becoming familiar with the general issues and concepts in educational measurement will enable you to understand the specific aspects of measurement that are introduced in later chapters.

## CHAPTER OUTLINE

The Role of Measurement in Education
       Do We Need So Much Measurement and Evaluation?
       Subjective Judgment, Objective Testing, and Quantification
Various Uses of Educational Measures
       Direct Instructional Decisions
       Instructional Management Decisions
       Entry-Exit Decisions
       Program Administrative and Policy Decisions
       Decisions Associated with Expanding Our Knowledge Base

## GUIDED STUDY EXERCISES

### Definition Exercises

Locate the following terms in your text and briefly define each in the context of educational measurement and evaluation.

Classification:

Professional Judgment:

Objective Measurement:

Quantification:

Snapshot Tests:

Diagnostic Decisions:

Prescriptive Tests:

Placement:

Selection:

Guidance Decisions:

**Multiple-Choice Questions**

Circle the letter of the item that best answers each question.

1. Recent data from testing company sales indicate that the use of tests is

   a. increasing.
   b. decreasing.
   c. remaining stable.
   d. fluctuating dramatically.

2. Approximately how many teacher-made tests does a typical classroom teacher give to students during a school year?

   a. 25
   b. 50
   c. 75
   d. 100

3. Which of the following statements is true concerning the concept of quantification?

   a. Some characteristics are harder to quantify than others.
   b. Quantification is central to the process of subjective measurement.
   c. Quantification is essential in using professional judgments.
   d. Most educational concepts or qualities cannot be quantified.

4. "Everyone who thinks the answer should be 220 please raise your hand." This statement is an example of which of the following types of tests?

   a. Guidance test
   b. Norm-referenced test
   c. Snapshot test
   d. Minimum competency test

5. For what types of decisions are tests used most frequently?

   a. Classification
   b. Selection
   c. Placement
   d. Diagnosis

6. Sarah has just taken a test designed to measure her vocational interests. The results of this test will most likely be used to make decisions about

   a. diagnosis.
   b. guidance. *for the counselor and Sarah, options*
   c. selection.
   d. certification.

   *admissions / acceptance*

7. Tests taken by persons wanting to graduate from high school, obtain a first aid certificate, or obtain a medical license are examples of

   a. guidance tests.
   b. curriculum-based tests.
   c. certification tests.
   d. classification tests.

   *confusing, typically no "one" test given to graduate from high school in U.S. v. European models*

8. Classification or placement decisions, and guidance decisions, are both examples of which of the following types of decisions?

   a. Entry-exit
   b. Instructional management
   c. Direct instruction
   d. Program, administrative, and policy

**Short-Answer Questions**

Provide a brief answer for each of the following questions in the space provided.

1. Although standardized testing has been widely criticized, there is evidence that the use of tests is increasing. Why is this so?

2. List three different educational functions for which test results play an important role.

3. Compare and contrast objective and subjective measurement.

4. Why is quantification important in measurement and evaluation?

5. List some ways that teacher-made tests are useful in making instructional planning decisions.

## APPLICATION ACTIVITIES

### To Test or Not to Test

The Hilldale School Board is holding one of its regularly scheduled public meetings. The particular topic being addressed at this meeting is how to increase student academic performance. Members of the community are concerned because average standard scores of students in the district have been slipping in comparison with neighboring school districts over the past three years. In response to such concerns, the school board has unanimously decided to adopt a more rigorous academic curriculum.

The issue being discussed at this meeting is how to assess the effectiveness of this new curriculum. Of particular interest is what type of testing system, if any, should be adopted to assess student performance. Two members of the audience provide the school board with strong but opposing positions about what should be done. Mr. Yelker is strongly opposed to the use of standardized tests. He advocates informal assessment of student progress, with a stronger emphasis on professional judgment by teachers. Ms. Lewen takes an opposing position. She strongly urges the school board to use standardized academic achievement tests to evaluate the effectiveness of the new curriculum.

Outline the pros and cons of using standardized tests versus informal assessment to evaluate the effectiveness of such a new curriculum. Compare your arguments with the information provided in the Application Activity Answers section of this chapter. Which position makes the most sense to you?

# ANSWERS TO GUIDED STUDY EXERCISES

## Definition Exercises

*Classification*: Categorizing students into groupings based on measurement results.

*Professional Judgment*: The process of evaluating student performance based on one's experience and intuition rather than through the use of formal, objective measures.

*Objective Measurement*: Collection of data in a way that most reasonable persons would interpret the data similarly.

*Quantification*: A process that underlies measurement and supports objectivity; it involves attempting to measure a behavior or characteristic in a manner that yields numerical, replicable information.

*Snapshot Tests*: Brief, informal oral questions used by teachers during a teaching activity.

*Diagnostic Decisions*: Educational decisions pertaining to the reasons why students are or are not learning.

*Prescriptive Tests*: Tests used to "prescribe" specific educational techniques or placements based on specific learning or performance styles; these tests suggest that different students should receive different treatments.

*Placement*: The process of dividing a population of students into different groups using instructional strategies or materials that are most appropriate for that group of students.

*Selection*: The process of selecting a limited number of individuals out of a larger group based on some specific criteria.

*Guidance Decisions*: Decisions involving planning for a student's future that might involve the use of interest tests, personal adjustment inventories, and other affective tests used by guidance counselors.

## Multiple-Choice Questions

1. A
2. D
3. A
4. C
5. D
6. B
7. C
8. B

## Short-Answer Questions: Key Points

1. Some of the reasons that standardized testing is increasing include (a) educational agencies and community groups want to use them to determine whether or not costly educational reforms are succeeding, and (b) there is an increasing public demand for objective evidence that teachers and schools are effectively educating students.

2. The broad educational functions in which test results play an important role are direct instructional decisions; instructional management decisions; entry-exit decisions; program, administrative, and policy decisions; and decisions associated with expanding our knowledge base.

3. Objective measurement is based on data and is also based on the premise that most reasonable persons confronted with the same data would interpret it in a similar fashion. Subjective measurement, on the other hand, is personal and peculiar to the individual making the judgment and is likely to vary considerably from person to person.

4. Quantification is an essential concept in evaluation and testing, as it supports objectivity and provides a basis for measurement.

5. Some of the ways that teacher-made tests are especially useful in making instructional planning decisions are (a) assessing students' performance on their actual curricula, (b) helping the teacher to carefully analyze instructional goals, and (c) directly assessing student performance over a period of time.

## APPLICATION ACTIVITY ANSWERS

### Arguments against the Use of Standardized Tests

Because of test anxiety and other factors, standardized tests do not fully reflect the potential performance of some students.

Standardized tests are subject to cultural, racial, and gender bias.

Tests do not measure some important information, which is reflected in teachers' professional judgments.

There is too much reliance on tests as a source of information for making policies.

### Arguments for the Use of Standardized Tests

Professional judgments are not always sound, because of subjectivity and bias.

Tests produce objective and quantifiable information that can be used to make judgments.

Tests provide standardized information about students' attainment of instructional goals.

Tests stimulate student effort, set performance expectations, and provide feedback on accomplishments.

Tests can be an effective way of confirming whether costly educational reforms have succeeded.

# CHAPTER 2
# HOW DID WE GET WHERE WE ARE IN MEASUREMENT?

## CHAPTER SUMMARY

George Santayana observed that those who cannot remember the past are condemned to repeat it. This may not be entirely true in the field of educational measurement and evaluation, but it is useful to understand how the field has evolved over time. This chapter traces some of the important historical events in measurement and evaluation and describes how the field has evolved to its present state.

The chapter begins with a discussion of some of the important "prescientific" uses of educational measurement. You may be surprised to learn that measurement and evaluation activities have been used at various times and places throughout history, including biblical times, and in ancient China and Greece. Next, the development of early scientific measurement techniques in the late 1800s and early 1900s is detailed, including the gradual shift to objective assessment and the development of the first intelligence tests. The period from 1920 to 1965 was characterized by the refinement and development of standardized testing practices and their ultimate widespread use. The final section of the chapter is devoted to a discussion of nine different recent developments and trends in educational measurement. This section ties the past to the present and describes some issues that may affect you directly. For example, you may have had to pass a competency exam to be admitted to the teacher education program at your college or university, or you may be required to pass a different competency exam before you can be certified as a teacher. And, if you ultimately become employed as a public school teacher, you will very probably have to deal with the issues of minimum competency testing for your students and whether or not the standardized achievement tests the students in your school take appropriately measure what they are being taught.

Study this chapter with an eye on the past and a look to the future, and remember that events in history can sometimes provide a good way of understanding what is happening in the present. You should pay special attention to the nine recent developments and trends in educational measurement, as these are things that may affect you directly--perhaps sooner than you think!

## CHAPTER OUTLINE

Early "Prescientific" Uses of Educational Measurement
Development of Early Scientific Measurement Techniques: Pre-1920
Trends in Educational Measurement: 1920 to 1965
Recent Developments and Trends in Educational Measurement
    National Assessment of Educational Progress
    The "Accountability" Movement
    The Trend toward Criterion-referenced Measurement
    Trends in Scholastic Aptitude and Achievement Test Scores

The Establishment of Minimum Competency Testing Programs
Renewed Calls for Reform in Educational Measurement
Development of Professional Organizations for Measurement Specialists
Positions Taken by Professional Associations toward Measurement Issues
The Use of Competency Tests for Teacher Certification
Calls for Alternative "Authentic" Performance Measures

## GUIDED STUDY EXERCISES

### Definition Exercises

Locate the following terms in your text, and briefly define each in the context of educational measurement and evaluation.

Recitation:


Elementary and Secondary Education Act (ESEA):


Educational Accountability:


Criterion-referenced Tests:


Minimum Competency Tests:


Standards for Educational and Psychological Testing:


Teacher Competency Tests:


National Council for Measurement in Education:

Stanford-Binet Intelligence Test:


"Prescientific" Measurement:


## Multiple-Choice Questions

Circle the letter of the item that best answers each question.

1. The oral questioning technique developed by Socrates provided the basis for which of the following educational techniques that were widely used in American schools well into the 1900s?

   a. The recitation method
   b. Formal lecture methods
   c. The lab and lecture method
   d. Guided discovery methods

2. Which statement regarding the National Assessment of Educational Progress is true?

   a. When introduced, NAEP was readily accepted by educators.
   b. Very few states have adopted testing programs based on the NAEP.
   c. It began in 1964 as a nationwide effort to assess academic performance of U.S. students.
   d. NAEP makes systematic comparisons of academic performance between states and districts.

3. Which of the following was the first published comprehensive achievement test battery in the United States?

   a. Iowa Tests of Basic Skills
   b. Carnegie Achievement Tests
   c. New York Board of Regents Exam
   d. Stanford Achievement Tests

4. The "accountability" in education

   a. began in the 1950s.
   b. contributed to the development of criterion-referenced testing.
   c. resulted in the ban of norm-referenced testing in many states.
   d. contributed to the discontinuation of minimum competency tests.

5. Since the mid-1980s, national average scores on tests of scholastic aptitude have

   a. declined.
   b. fluctuated dramatically.
   c. increased.
   d. remained relatively stable.

6. The use of "minimum competency tests" has

   a. been rejected in most states.
   b. been found to be unconstitutional.
   c. never been challenged in U.S. courts.
   d. been criticized as being politically motivated.

7. Which professional organization for educators voted for a moratorium on standardized testing in the schools in the 1970s?

   a. National Education Association
   b. American Federation of Teachers
   c. National Association of Secondary Principals
   d. National School Boards Association

8. Which of the following statements about the use of minimum competency testing for teachers is *not* true?

   a. A majority of Americans favor its use.
   b. A majority of states test or plan to test applicants to teacher training programs.
   c. A majority of states test or plan to test applicants for teacher certification.
   d. A majority of states test or plan to test teachers for recertification.

## Short-Answer Questions

Provide a brief answer for each of the following questions in the space provided.

1. Describe the events, persons, and findings that led to the widespread use of written tests as an alternative to the use of oral tests in the late 1800s and early 1900s.

2. List some of the problems with written essay examinations that led to the development and refinement of objective testing.

3. How do the objectives and procedures of criterion-referenced testing differ from those of norm-referenced testing?

4. What are some of the criticisms of the use of minimum competency tests for making pass/fail decisions about student performance in educational programs?

5. What are some of the arguments in favor of and against the use of competency tests to license teachers to work in the public schools?

## APPLICATION ACTIVITIES

The use of standardized, norm-referenced tests to measure academic achievement in the schools has been roundly criticized on many grounds, in spite of their widespread use and the advantages they offer. Assume for a minute that you are a teacher employed in a public school district that has just placed a moratorium on all districtwide standardized testing used to measure student achievement and make placement decisions. You have been given the responsibility of developing an alternative "authentic" performance test for your content area/specialization area (if you are not training to become a teacher, select a subject matter or specialty area that interests you). Given this charge, do the following:

1. Develop and describe an alternative assessment method and a rationale for it.

2. Provide a specific example of how you would use this alternative assessment method to measure student performance against instructional objectives in your area.

## ANSWERS TO GUIDED STUDY EXERCISES

### Definition Exercises

*Recitation*: A mode of oral questioning that was commonly used in American schools well into the 1900s and was roughly based on the questioning method developed by Socrates.

*Elementary and Secondary Education Act*: A federal program initiated in 1965 that mandated evaluation studies in thousands of public schools and spurred the development of testing technology.

*Educational Accountability*: A concept based on a movement that was popular in the 1970s, wherein school personnel were held liable for educating students.

*Criterion-referenced Tests*: An alternative to norm-referenced testing that bases test items on what is taught in the curriculum and links them to instructional targets.

*Minimum Competency Tests*: Objective tests mandated by state or local governing bodies that require attainment of minimum standards of competency for educational pass/fail decisions.

*Standards for Educational and Psychological Testing*: A set of professional and ethical standards for test development and use that have been jointly developed by three professional associations concerned with the use of tests.

*Teacher Competency Tests*: Tests adopted in some states that must be completed at an acceptable level to receive a teaching certificate.

*National Council for Measurement in Education*: A professional organization whose goal is the advancement of sound measurement techniques in education.

*Stanford-Binet Intelligence Test*: The first intelligence test published in the United States, which was adapted by Lewis Terman in 1916 from Alfred Binet's earlier intelligence test from France.

*"Prescientific" Measurement*: Methods of testing and measurement used prior to the development of scientific measurement techniques.

## Multiple-Choice Questions

1. A
2. C
3. D
4. B
5. D
6. D
7. A
8. D

## Short-Answer Questions: Key Points

1. Several individuals and events were prominent in the increased use of written tests as an alternative to oral tests. Educational reformer Joseph Bice called attention to the widely varying conditions tests were administered under and developed and administered written standardized tests. Psychologist and researcher E. L. Thorndike was influential in persuading educators to adopt precise measurement techniques such as standardized written tests. The development of the first standardized intelligence test by Alfred Binet in France was also very influential in shaping the use of standardized tests.

2. The use of written essay examinations was found to be suspect through a series of studies in the early 1900s that showed wide variation in the way grades were assigned, even when the same written papers were graded by the same evaluators on different occasions. These findings provided ammunition for advocates of standardized, objective testing such as E. L. Thorndike.

3. Criterion-referenced testing differs from norm-referenced testing in that it attempts to measure student performance against an objective set of criteria. CRTs tend to provide better instructional targets than NRTs, as well as a better match between what is taught and what is tested. NRTs, on the other hand, provide information on how a student's test performance compares with other students from a specific normative group.

4. The use of minimum competency tests for making pass/fail educational decisions has been criticized on grounds that such use is more politically motivated than educationally sound and reflects blind faith in governmental bodies attempting to ensure minimum educational standards. Another criticism of using MCTs for pass/fail educational decisions is that the actual minimum performance standard required for passing is usually arbitrary and may not reflect a relevant level of attainment.

5. Arguments in favor of using MCTs for teacher certification include public support for the concept, the notion that both schools and society are best served if teacher competence is assured, and the general tenets of the educational accountability movement. Arguments against using of MCTs for this purpose include the questionable validity of the MCTs for ensuring that teachers are competent and the fact that these tests tend to focus only on basic literacy and do not ensure competency on other criteria that are important in teaching.

# APPLICATION ACTIVITY ANSWERS

Since the specific content area/specialization area you have selected and the way you approach the task will vary from that of other students, it is not possible to provide one specific "correct" answer. However, the following example is provided as a general guide to how an alternative assessment method could be used. The subject or content area for this assessment approach is primary or elementary-level reading.

1. <u>Method and Rationale</u>. The alternative assessment method selected for measuring elementary or primary-level reading skills is *curriculum-based measurement* (CBM). This method is based on assessing reading performance as they actually read from their assigned materials. The rationale behind it is that assessing students' performance on their actual curriculum is a very useful measurement procedure--it directly links the students' curriculum with the test, since they are essentially the same.

2. <u>Example for Measuring Student Performance against Instructional Objectives</u>. Assuming that a teacher had developed an instructional objective that students will be able to read the material in reading unit 4 at the rate of 30 words per minute at 90 percent accuracy, measuring student performance against this objective is relatively easy. The student will read orally from reading unit 4 for exactly three minutes, while the teacher follows along on the same reading material and marks any errors that occur. The three minutes are divided into three one-minute units, and the teacher counts how many words the student correctly reads and the overall accuracy rate for each one-minute period. Finally, an average is calculated for the three one-minute reading samples (for example, 34 words correctly read per minute, at an 88 percent accuracy rate), and these figures are compared against the instructional objectives. Mastery of the instructional objectives would indicate that the student is ready to move on to the next unit; whereas failure to master the criteria would indicate that the student is in need of additional remedial reading instruction and practice.

# CHAPTER 3
# COMING TO GRIPS WITH CURRENT SOCIAL, LEGAL, AND ETHICAL ISSUES IN MEASUREMENT

## CHAPTER SUMMARY

Thus far, several current, and sometimes controversial, issues relating to measurement and evaluation have been introduced: test bias, the use of minimum competency tests, privacy and confidentiality in testing, legal regulation of testing practices, and the use of tests for teacher certification. Chapter 3 will help you explore each of these issues in greater depth and will allow you to develop a basic understanding of acceptable and unacceptable practices in each area.

The chapter begins with a discussion of social and legal considerations in testing, and this discussion is dominated by the issue of test bias in its various forms, specifically including cultural, socioeconomic, and gender bias. This discussion of test bias includes a review of some of the landmark court decisions regarding the use of tests. Next some issues in minimum competency testing (MCT) are explored, including a review of some important legal decisions affecting the use of MCTs and a discussion of which practices with MCTs are thought to be legally defensible and which are not. Following the section on MCTs, right to privacy and test disclosure issues are discussed, followed by a closely related topic, ethical considerations in using educational measures. Finally, the current state of the use of tests for teacher certification is covered, and the chapter ends with an exploration of how in spite of the numerous legal challenges testing practices have faced, there seems to be a renewed and increased demand for tests that is not likely to diminish in the near future.

You may have heard the old adage "ignorance of the law is no excuse." In the field of educational measurement, there is truth to this saying, and it could be expanded to "ignorance of legal and ethical considerations in testing is no excuse for violating them." There have been numerous legal cases where education and psychology professionals have been held liable for violating legal and ethical codes of behavior relating to the use of tests, and ignorance of legal and ethical standards has not gotten any of them "off the hook." Study this chapter carefully and use the information in it as a guide to "best practices" and "things to avoid" in your use of tests.

## CHAPTER OUTLINE

Social and Legal Considerations in Testing
    Test Bias and Discrimination Issues
        Concerns about Cultural, Ethnic, and Linguistic Test Bias
        Concerns about Socioeconomic Bias in Tests
        Concerns about Gender Bias in Tests
Issues in Minimum Competency Testing
    Standard Setting

    Legal Challenges to MCTs
    Summary of What the Courts Have Said about MCTs
Right to Privacy Issues in Testing
Test Disclosure Issues in Testing
Ethical Considerations in Using Educational Measures
    Ethical Responsibilities of Measurement Professionals
    Ethical Responsibilities of Test Users
    Ethical Problems Associated with Teaching to the Test
Issues in Using Tests for Teacher Certification

## GUIDED STUDY EXERCISES

### Definition Exercises

Locate the following terms in your text and briefly define each in the context of educational measurement and evaluation.

Test Bias:


Minimum Competency Testing:


Representative Norm Group:


Test Disclosure:


Privacy Rights:


Teacher Competency Testing:


FERPA:

## Multiple-Choice Questions

Circle the letter of the item that best answers each question.

1. Which of the following factors is *not* known to potentially produce bias in test results?

   a. Methods of test construction
   b. Methods of test selection
   c. Methods of determining test reliability
   d. Methods of test administration

2. Which of the following provides evidence that a test is biased?

   a. One individual or group has an unfair advantage over another.
   b. The results are not made available through test disclosure.
   c. The test is written in only one language.
   d. The test measures intellectual ability rather than academic achievement.

3. Which of the following explanations for significant differences in ability test scores between whites and minority group members in the United States is the most plausible?

   a. Minorities may have been denied opportunities to develop skills measured on the tests.
   b. Most IQ tests used to measure intellectual ability are poorly constructed.
   c. There is strong scientific evidence for racially based genetic differences in intelligence.
   d. Test developers inadvertently introduce cultural bias into ability tests.

4. Which of the following statements regarding the 1954 U.S. Supreme Court case *Brown v. Board of Education of Topeka* is true?

   a. The case was later overturned as unconstitutional.
   b. The decision upheld the use of tests with racial minorities.
   c. The court challenged special education placement procedures for EMR classes.
   d. The right of all children to equal educational opportunities was established.

5. An achievement test that mentions female characters more often than male characters would be

   a. a realistic portrayal of cultural attitudes.
   b. selectively biased against males in content.
   c. selectively biased against females in language.
   d. more difficult for male subjects.

6. Which of the following examples would be the area where a minimum competency test would be most likely to be found illegal?

   a. Students with disabilities are denied a high school diploma if they do not pass the test.
   b. Some 20 percent of Hispanic students but only 1 percent of Asian students are denied high school diplomas if they do not pass the test.
   c. There is very little match between instructional programs and the test.
   d. Successfully passing the test requires a minimum of a seventh grade reading level.

7. Which of the following statements about the Family Education Rights and Privacy Act of 1974 (FERPA) is true?

   a. The law covers educational institutions only in the states of New York and California.
   b. It has made only a minor impact on the field of educational testing.
   c. The law made it illegal to use minimum competency tests for teacher certification.
   d. It placed test disclosure regulations on schools that receive federal monies.

8. Which of the following would *not* be considered an ethical responsibility of test users?

   a. They should have a general understanding of measurement principles.
   b. They should have had some graduate-level coursework in measurement and evaluation.
   c. They should be knowledgeable about the particular test and its uses.
   d. They should receive training in using the test.

**Short-Answer Questions**

Provide a brief answer for each of the following questions in the space provided.

1. List the characteristics of legally defensible minimum competency tests, and also list characteristics that would make these tests subject to legal action.

2. Contrast the court ruling in the following two cases regarding the use of IQ tests with minority group members: *Larry P. v. Wilson Riles*, and *PASE v. Hannon*.

3. Provide two examples of gender-based test bias.

4. Provide two examples of instances where it would be ethically permissible to violate the confidentiality of test results.

5. Given that the use of tests has been so vigorously challenged in court, is it likely that test use will decline in the future? Why?

## APPLICATION ACTIVITIES

### Minimum Competency Tests and the Law

Chapter 3 provides a very detailed overview of the legal considerations of using minimum competency tests (MCTs), which have been vigorously challenged in the courts. Since the use of MCTs remains a controversial issue, and since they are being used in public school districts at an increasing rate, it is to your advantage to develop a basic working knowledge of what uses of MCTs are and are not legally defensible at this point.

The following list includes examples of MCT use that are likely to run afoul of the law and examples that have been shown to be legally defensible. Review the "Issues in Minimum Competency Testing" section in Chapter 3, and then place the number corresponding to each of the following items into one of these two categories: "Legally Questionable," or "Legally Defensible." Use the table on the next page to record your answers.

1. Use of an MCT results in 22 percent of minority group members at a high school being denied diplomas, while only 3 percent of white students are denied.

2. An MCT requires all students to perform at the 8.5 grade level in reading, mathematics, and written language, but this criterion results in boys being denied high school graduation at a significantly higher rate than girls.

3. An MCT used as a basis for grade promotion has been found to have content and standardization properties that bias it against certain minority groups.

4. A school district requires students with disabilities to score at the same level on MCTs as other students in order to receive a diploma.

5. A school district requires students with disabilities to score at the same level on MCTs as other students to receive a diploma, even though students with certain physical disabilities are not able to adequately demonstrate their actual knowledge on these tests.

6. A school district requires specific levels of MCT performance for promotion to the next grade. The tests are not biased in content or standardization but have questionable psychometric properties such as reliability and validity.

7. A school district decides to raise the minimum standards required on MCTs for graduation from high school. This change in standards results in certain racial or ethnic group members being denied diplomas at a disproportionally high rate.

8. With no advance notice, a school district adopts an MCT battery to be used for grade promotion and graduation at the end of the school year.

| Classification of Eight Items Related to the Use of Minimum Competency Tests ||
| LEGALLY QUESTIONABLE | LEGALLY DEFENSIBLE |
| --- | --- |
| Items: | Items: |

## ANSWERS TO GUIDED STUDY EXERCISES

### Definition Exercises

*Test Bias*: Factors that provide an unfair advantage on a test for certain individuals or groups.

*Minimum Competency Testing*: Tests mandated by state or local governing bodies that require attainment of minimum standards of competency for educational pass-fail decisions.

*Representative Norm Group*: Standardization groups for norm-referenced tests that are representative of the population as a whole on such factors as race or ethnicity, socioeconomic status, and gender.

*Test Disclosure*: Requirements that test developers publicly disclose the contents of certain tests that are used to make important decisions about the examinees' futures.

*Privacy Rights*: Individual rights to privacy and confidentiality that may affect the testing process, and the need to obtain informed consent when such rights are waived.

*Teacher Competency Testing*: A form of minimum competency testing that requires teachers to pass various basic skills tests prior to receiving or renewing their teaching certificate.

*FERPA*: The Family Educational Rights and Privacy Act of 1974, also referred to as the "Buckley Amendment," which requires all educational institutions receiving federal funds to allow parents and eligible students access to records pertaining directly to the students' school performance.

## Multiple-Choice Questions

1. C
2. A
3. A
4. D
5. B
6. C
7. D
8. B

## Short-Answer Questions: Key Points

1. Legally defensible MCTs are technically sound, nonbiased, and closely matched to instructional programs they are intended to be used with. MCTs would be subject to legal action if they had potential for discrimination based on race or language, if insufficient advance notice was given before they were phased in, if they were technically unsound, if they did not match the instructional program, or if they created or reinforced tracking systems.

2. Both the *Larry P.* and *PASE* cases involved the use of IQ tests for special education placement. In the *Larry P.* decision, it was ruled that the use of IQ tests was biased and discriminatory against African-American students. In the *PASE* decision, the judge, after examining the content of each test item, ruled that very few items were culturally biased and that IQ tests did not discriminate against black children.

3. Examples of gender bias in tests: Test items containing numerous sex-role stereotypes, mentioning one gender more frequently, or having an imbalance of male to female nouns and pronouns.

4. The following circumstances reflect instances when it would be ethically permissible to violate the confidentiality of test results: (a) The results reveal that there is a clear and immediate danger to the students or others, (b) sharing the results with other professionals would significantly help the student, (c) the test results are used to make professional decisions about the student and must therefore be shared with other professionals, and (d) the student waives the right to confidentiality.

5. Although the use of tests has been vigorously challenged, it is unlikely that test use will decline in the near future. One reason is that there are few alternatives for measuring student ability and achievement. Another reason that a decline is unlikely is that there are continuing calls for increased accountability by teachers and schools for the education of students.

## APPLICATION ACTIVITY ANSWERS

### Minimum Competency Tests and the Law

The most appropriate categorization ("legally questionable" versus "legally defensible") for the eight examples related to minimum competency testing is as follows:

| Classification of Eight Items Related to the Use of Minimum Competency Tests ||
|---|---|
| LEGALLY QUESTIONABLE | LEGALLY DEFENSIBLE |
| Items 3, 5, 6, and 8 | Items 1, 2, 4, and 7 |

# CHAPTER 4
# GETTING YOUR BEARINGS: SOME CONCEPTS AND CLASSIFICATIONS BASIC TO MEASUREMENT AND EVALUATION

## CHAPTER SUMMARY

The first three chapters of your text have made many references to some of the basic concepts, classifications, and terminology in measurement and evaluation, but none have yet been treated in considerable detail. Chapter 4 builds on the base of knowledge you have thus far obtained in the area of educational measurement and evaluation by covering some of the basic concepts of the field in considerable detail. This chapter will provide you with a solid foundation for expanding the technical knowledge and practical skills you will acquire in the coming chapters.

The chapter begins with some key definitions and distinctions. You are already familiar with some of the terms, such as *reliability, validity, objectivity, evaluation,* and *testing,* and others are defined and illustrated in practical detail. Next, a detailed outline of how we classify measurement instruments is presented, with 13 different classification categories. Finally, you are exposed to an outline of how educational measurement can be roughly divided into five broad categories. Here you will learn such distinctions as how data collected from a student diary differ in nature from data collected by a videocassette recorder.

In later chapters you will receive information, examples, and exercises that will help you do such things as effectively develop your own tests, develop a good scoring system, and interpret the data you collect. Developing a solid understanding of the basic concepts and terminology in educational measurement and evaluation from this chapter will prove to be a necessary step to take before you jump into these other activities.

## CHAPTER OUTLINE

Definitions, Distinctions, and Terminology
    Some Simple Definitions
    Distinguishing among Testing, Measurement, Assessment, and Evaluation
Mapping the Terrain: Classification of Measurement Instruments
    By Discipline or Subject Area of the Content
    By Psychological Constructs
    By Specific Skills, Type of Knowledge, or Behaviors Measured
    By Item Format
    By the Type of Data Produced
    By the Purpose for Which They Are Used
    By How the Instrument Is Administered

By How the Instrument Is Scored
By Type of Performance Called For
By Ways of Recording Behavior
By the Types of Objectives Measured
By How the Score Is Interpreted
By Who Constructs Them, and How
One Measure May Fit Several Classifications
An Outline of Types of Educational Measurement Instruments
    I.    Data Recorded by and Collected from Individuals
        A. Self-Reports
        B. Personal Products
    II.   Data Recorded by an Independent Observer
        A. Written Accounts
        B. Observation Forms
    III.  Other Data Collection Procedures
        A. Mechanical Devices
        B. Unobtrusive Measures
        C. Existing Information Resources or Repositories

## GUIDED STUDY EXERCISES

### Definition Exercises

Locate the following terms in your text and briefly define each in the context of educational measurement and evaluation.

Test:

Measurement:

Objectivity:

Reliability:

Validity:

Assessment:

Evaluation:

Norm-referenced Test:

Criterion-referenced Test:

Objectives-referenced Test:

Domain-referenced Test:

Standardized Test:

Teacher-Made Test:

**Multiple-Choice Questions**

Circle the letter of the item that best answers each question.

1. Tara has taken a standardized aptitude test at the beginning of the school year for the past three years. Her scores (based on an average score of 100) have been 103, 101, and 102. Based on this information alone, which of the following psychometric properties does this test demonstrate?

    a. Content validity
    b. Criterion-related validity
    c. Reliability
    d. Objectivity

2. The Central School District has just completed a three-month process using a variety of procedures to determine the quality and effectiveness of the elementary math curriculum. Which of the following procedures best describes what was done?

    a. Evaluation
    b. Accountability
    c. Testing
    d. Measurement

3. Which of the following descriptions are the terms *creativity, intelligence, attitude,* and *achievement* all examples of?

    a. Subject areas within a discipline
    b. Psychological constructs
    c. Specific skill areas
    d. Score interpretations

4. Kali, a fifth grade student, has just taken an academic achievement test that covers reading, written language, and mathematics. Her teacher scores the test and finds that Kali's overall score level is at the 72nd percentile level based on all fifth grade students in the district. What type of test is this?

    a. Criterion-referenced
    b. Objectives-referenced
    c. Domain-referenced
    d. Norm-referenced

5. Which of the following statements is true regarding nonstandardized teacher-made tests?

   a. They are administered more often than standardized tests.
   b. They are administered by the average teacher about 40 times per year.
   c. They tend to be less essential for instructional purposes than standardized tests.
   d. In terms of classroom measurement and evaluation, they are not as essential.

6. Which of the following is *not* an example of a self-report measures?

   a. Interviews
   b. Questionnaires
   c. Behavioral observations
   d. Checklists and inventories

7. A work sample would belong to which of the following categories of measures?

   a. Data recorded by an independent observer
   b. Data collected by a mechanical device
   c. Self-reports
   d. Personal products

8. Sociometric measures are to self-reports as a review of public documents is to

   a. personal products.
   b. data recorded by an independent observer.
   c. data collected from existing information resources.
   d. rating scales.

## Short-Answer Questions

Provide a brief answer for each of the following questions in the space provided.

1. Compare and contrast speed tests and power tests.

2. What are some ways of conducting unobtrusive measurement in schools?

3. How do CRTs, ORTs, and DRTs overlap, or, in other words, how are these categories similar?

4. Compare and contrast the terms *objectivity* and *subjectivity*.

5. List the four most common types of supplied item tests.

## APPLICATION ACTIVITIES

### Learning to Differentiate Between NRTs, CRTs, ORTs, and DRTs

Chapter 4 introduced you to a classification of tests based on how the score is interpreted. Four different test categories were introduced, including norm-referenced tests (NRT), criterion-referenced tests (CRT), objectives-referenced tests (ORT), and domain-referenced tests (DRT). This breakdown is probably still a little bit vague for you, but it is important to learn the terminology and concepts involved here, as classroom teachers and other education professionals are likely to work with these different test types on a regular basis.

First, review the section on classifying educational measures by how the score is interpreted, paying close attention to the four examples provided in boxes. Then, carefully read the following four examples, and decide which of the four categories (NRT, CRT, ORT, DRT) each belongs to. You can check your work by referring to the information in the Application Activity Answers section of this chapter.

**Example 1.** The Mapletown School District has recently adopted a minimum competency test that must be passed by high school seniors before they are awarded a diploma. This test covers basic general information from each academic area from which all high school students are required to take some coursework, such as general science, English, U.S. history, and math. Scores are simply reported as percentages of the test items that are correctly answered. In order to pass the test and receive a diploma, students are required to get a minimum of 75 percent correct on each of the area tests and a minimum of 80 percent correct on the total score.

Example 1 is a description of a _____.

**Example 2.** In addition to the test required by high school seniors in the Mapletown School District that was described in Example 1, the district also conducts districtwide academic achievement assessment each spring at every grade level. These test scores are not used to make minimum competence decisions, but to provide teachers and parents with information about how individual students are doing in comparison with other students nationwide and within the district. The information is also used for the school board and administration to determine how well their students are doing in comparison with other school districts nationwide. For each student, class, grade level, and for the district as a whole, individual and average scores are given in percentile ranks and standard scores.

Example 2 is a description of a _____.

**Example 3.** Leon and Katherine both teach basic science courses at a middle school. At the end of the first semester basic science unit, these two teachers get together and develop a pool of 250 questions that they believe covers the broad content of basic seventh grade science, according to what they teach. They randomly select 40 questions out of the pool of 250 and agree to administer them to their students to measure their level of achievement. Leon and Katherine have estimated that if they randomly selected any other 40 items out of the 250-item pool, the test would be equally difficult and would measure knowledge from the same general areas.

Example 3 is a description of a _____.

**Example 4.** Chris, a second grade teacher at Washington Elementary School, is finishing up a mathematics unit with his students and wants to know how well they have mastered the goals he established at the beginning of the unit. He creates a math test with 40 items. On his scoring key, he identifies not only what the correct answer should be but which of the following four groups each item belongs to: (a) single-digit addition, (b) two-digit addition, (c) single-digit subtraction, and (d) two-digit subtraction. Chris' goal at the beginning of the unit was for each student to be able to complete problems in each of the four areas with 80% accuracy. The test he has developed will allow him to determine whether or not each student has met that instructional goal.

Example 4 is a description of a _____.

## ANSWERS TO GUIDED STUDY EXERCISES

### Definition Exercises

*Test*: One form of measurement; a set of tasks or questions presented to an examinee in a systematic manner, used to elicit a sample of behavior on the attribute or characteristic of interest.

*Measurement*: The process of making empirical observations of something and translating those observations into quantifiable or categorical form according to specific rules and procedures.

*Objectivity*: The degree to which two or more reasonable persons who are given a scoring key or scoring criteria would agree on how to score each test item and the number of points each examinee should receive for the total test.

*Reliability*: How consistently a test measures something.

*Validity*: The degree to which a test measures what it was intended to measure.

*Assessment*: The use of specially constructed tests to determine the overall condition of an entire system or its major components.

*Evaluation*: The determination of a thing's worth, value, or quality.

*Norm-referenced Test*: A test that shows how a student's performance stands in relation to other students.

*Criterion-referenced Test*: A test that shows how a student's performance compares with some absolute standard or criterion of success.

*Objectives-referenced Test*: A test that provides information about how well the student performs on test items measuring attainment of specific instructional objectives for which the test was developed.

*Domain-referenced Test*: A test consisting of a random sample of items drawn from a pool representative of all test items for a well-defined content area.

*Standardized Test*: A set of tasks presented to individuals under similar (standard) conditions and designed to measure some particular aspect of those individuals' behavior and allowing individuals' scores to be arrayed on a measurement scale for that behavior.

*Teacher-Made Test*: A test constructed entirely by a teacher for the purpose of testing pupils in that teacher's classroom(s).

## Multiple-Choice Questions

1. C
2. A
3. B
4. D
5. A
6. C
7. D
8. C

## Short-Answer Questions: Key Points

1. Speed tests are designed to be administered within a specified time frame, in order to measure how well students perform on the test within a given length of time. Power tests, on the other hand, generally do not have strict time limits but are designed to measure how well the students perform on the test items, regardless of time factors.

2. Unobtrusive educational measures involve gathering data in a manner so that others are not aware. Examples of unobtrusive measures include existing file materials or work samples and existing information such as library records, the condition of educational materials, the content of graffiti on the walls, and so forth.

3. CRTs, ORTs, and DRTs are all thought to be variant forms of criterion-referenced measurement, in that they are different ways of assessing how a student is performing according to a specific educational criterion.

4. Objectivity refers to the degree to which two or more reasonable persons, given the same scoring criteria, would agree on how to score each test item, resulting in the scorers assigning the same number of points for the same item and arriving at the same total score. Subjectivity is the degree to which scorers cannot agree on how to grade each test item, resulting in scorers assigning differing numbers of points to the same item and arriving at different total scores for the same student. In educational measurement, objectivity is to be strived for and subjectivity avoided to the greatest extent possible.

5. The four most common types of supplied item tests include essay, completion items, short-response items, and problem-solving items.

## APPLICATION ACTIVITY ANSWERS

Example 1 is a description of a *criterion-referenced test*.

Example 2 is a description of a *norm-referenced test*.

Example 3 is a description of a *domain-referenced test*.

Example 4 is a description of an *objectives-referenced test*.

# CHAPTER 5
# LEARNING TO READ THE SIGN POSTS: WHAT DO THOSE TEST SCORES REALLY MEAN?

## CHAPTER SUMMARY

Chapter 5 of your text gets you firmly into some of the "nitty-gritty" technical aspects of measurements related to scoring and interpreting test scores. Looking at the different technical names and formulas may seem intimidating to you at first, but don't panic--the chapter has been written in a "user friendly" manner and is designed to remove some of the anxiety that we often associate with number crunching.

The chapter is divided into two major sections. The first section provides an overview of the basic tools of score interpretation, along with plenty of practical examples and some exercises. Some of the specific measurement tools that you will learn about include score distributions, measures of central tendency, measures of dispersion, the concept of correlation, and two measurement techniques that relate to error in testing--confidence intervals and standard error of measurement. The major section of the chapter covers different ways of interpreting data. You are introduced to how test norms work, what kinds of scores standardized tests yield, interpreting score profiles, and ways of understanding relationships between different types of scores.

Take a deep breath, relax, and jump into learning about some of the basic principles of test and measurement statistics. In today's classrooms, teachers are confronted with a wide variety of measurement data that is gathered on their students, in addition to being responsible for developing their own informal measurement and evaluation systems. The fear and trepidation some educators feel about learning measurement and evaluation statistics is usually more hype than it needs to be. This chapter and the associated exercises and activities will help you become firmly grounded with a basic working knowledge of the things you really need to know.

## CHAPTER OUTLINE

Acquiring the Basic Tools of Interpretation
    Organizing Measurement and Evaluation Data
    The Normal Curve and Other Distributions
    Measures of Central Tendency
    Measures of Dispersion
    The Correlation Coefficient: A Measure of Association
    Measurement Error
Interpreting Test Scores and Data
    Norms
    Scores Yielded by Standardized Tests
    Profiles
    Relationship among Types of Scores

## GUIDED STUDY EXERCISES

### Definition Exercises

Locate the following terms in your text and briefly define each in the context of educational measurement and evaluation.

Normal Curve:


Skewed Distribution:


Measures of Central Tendency:


Measures of Dispersion:


Correlation Coefficient:


Confidence Interval:


Objective Mastery Score:


Standard Error of Measurement:


Test Norms:

Grade Equivalent Score:


Percentile Score:


Z-score:


Normal Curve Equivalent Score:


Stanines:




**Multiple-Choice Questions**

Circle the letter of the item that best answers each question.

1.  Which of the following is best used to show continuous data such as test scores?

    a. Histogram
    b. Frequency polygon
    c. Normal curve
    d. Standard deviation

2.  Which of the following score distributions or "curves" is said to be "bell shaped"?

    a. Normal
    b. Bimodal
    c. Skewed
    d. Asymmetrical

3. Which one of the following is the best indicator of central tendency when you want to minimize the influence of extreme test scores in a distribution?

   a. mean
   b. median ✓
   c. mode

4. On an algebra test, Maria's score was at the 82nd percentile. [rank] Which of the following statements is true? pg 125

   a. Maria's score was lower than 82 percent of the students who took the test.
   b. Eighteen percent of the students taking the test scored lower than Maria.
   c. Eighty-two percent of the students taking the test scored lower than Maria. ✓
   d. Maria answered 82 percent of the test questions correctly.

5. A statistically significant correlation implies

   a. an important relationship between two things.
   b. that the relationship between two things is strong enough that it most likely did not occur by chance. ✓
   c. a cause and effect relationship between two things.
   d. that the relationship between two things is so important that it cannot be explained away.

6. Which of the following statements correctly describes the relationship between observed scores, true scores, and error scores?

   a. observed score = true score + error score ✓
   b. observed score + true score = error score
   c. error score - true score = observed score
   d. observed score - true score = error score

   pg 143  $X = T + E$  and pg 118

7. Which of the following would be most likely to be shaped like a "normal" distribution of scores?

   a. The distribution of body weight of a group of professional wrestlers
   b. The distribution of scores for a standardized academic achievement test that was administered to 5,000 randomly selected elementary school students
   c. The distribution of GED test scores from all persons who failed the test in one year
   d. The distribution of scores of all seventh grade students at a school who scored above the 75th percentile on a math test

8. Michael recently took a standardized aptitude test. One of the scores that was reported was -1.25. Which type of score is this most likely to be?

   a. Objective mastery score
   b. Normal curve equivalent
   c. Z-score
   d. Percentile score

## Short-Answer Questions

Provide a brief answer for each of the following questions in the space provided.

1. Compare and contrast the information that is provided by grade equivalent scores and percentile scores.

2. When is each of the three measures of central tendency (mean, median, mode) best used?

3. Provide an example or make a drawing of each of the following types of score distributions: normal, skewed, and bimodal.

4. Professor Wilson recently discovered that low scores on academic achievement tests are correlated with high rates of behavioral problems. Why is it a mistake to assume that one causes the other?

5. Why is the use of confidence intervals important in interpreting test scores?

# APPLICATION ACTIVITIES

## Manipulating and Interpreting a Real Data Set

Over the course of the previous eight months, Aria, a special education teacher at Chavez Elementary School, has referred several students with learning problems to the school district psychologist. The purpose of the referrals was to have a comprehensive psychological and educational test battery administered to them in order to identify the particular strengths, weaknesses, and learning patterns of the students. One of the tests that the psychologist administered to a number of students was the Wechsler Intelligence Scale for Children, 3rd Edition (WISC-III). This is a comprehensive individual intelligence test that has an average score of 100. One day, Aria decides to open the files of some of the referred children and make comparisons. She finds 15 students that have taken the WISC-III, and their scores are as follows:

| | | |
|---|---|---|
| 82 | 102 | 111 |
| 91 | 112 | 91 |
| 72 | 98 | 76 |
| 98 | 91 | 84 |
| 85 | 79 | 87 |

Using this data set of 15 WISC-III scores, help Aria answer the following questions:

1. What are the three measures of central tendency for this data set (mean, median, mode)?

2. What is the range?

3. Based on the average score of 100 for the WISC-III, what type of distribution is this?

4. Based on the measures of central tendency, what can you say about the average performance of these students on the WISC-III?

# ANSWERS TO GUIDED STUDY EXERCISES

## Definition Exercises

*Normal Curve*: A bell-shaped symmetrical distribution of scores according to which many human characteristics are distributed.

*Skewed Distribution*: A score distribution where the majority of data points are clustered at one end of the distribution; the most common departure from the normal curve.

*Measures of Central Tendency*: Various ways of showing the middle or average range of test scores, including mean, median, and mode.

*Measures of Dispersion*: Measures that provide information on variability and differences of scores, including range and standard deviation.

*Correlation Coefficient*: A measure that can range from -1.0 to +1.0 that shows how two different things are related to each other. Correlation can be used to predict but not to show causation.

*Confidence Interval*: The range within which you can be fairly confident that a true score lies.

*Objective Mastery Score*: A score system that provides specific information on the type of test items passed or failed, as well as comparing the student's test scores with those of a norm group.

*Standard Error of Measurement*: A figure related to the concept of test reliability that provides a uniform way of estimating the degree of error in a test. [derived from the standard deviation and the reliability coefficient]

*Test Norms*: The data from a test that are gathered by administering it to a group of people who serve as a comparison or reference for individual scores. [referent group]

*Grade Equivalent Score*: A type of score yielded by standardized tests that tells what grade level an individual's score is equal to for the average score for students.

*Percentile Score*: A score indicating the percentage of students who score below a given point in the norm reference group.

*Z-score*: The distance between a particular score and the average score (mean) in standard deviation units.

*Normal Curve Equivalent Score*: A test score system with a mean of 10 and standard deviation of 21.06 that corresponds to percentile scores at the 1st, 50th, and 99th percentiles.

*Stanines*: A test score system that divides the normal distribution into nine equal units, each 1/2 standard deviation wide. Stanine scores range from 1 to 9.

## Multiple-Choice Questions

1. B
2. A
3. B
4. C
5. B
6. A
7. B
8. C

## Short-Answer Questions: Key Points

1. Grade equivalent scores provide an estimate of what grade level a score is equivalent to the average score of. Percentile scores give an indication of how a score ranks in comparison with a group of scores, by providing a percentage of scores in the reference group that fall below the level of that score. Both types of scores provide some information about the level of a given score, but in different ways; percentile scores provide a ranked comparison within a group of scores, while grade equivalent scores provide an estimate of a functioning level.

2. The mean is the best measure of central tendency when an exact average score is wanted and the distribution of scores is relatively normal. The median is the best measure when you want to determine the exact midpoint in a distribution, or when the distribution is not normally shaped. The mode is the best measure when you are interested in determining scores obtained by the largest number of persons. It is interesting to keep in mind that on a perfectly bell-shaped normal curve, the three measures of central tendency would all be in the same place.

3. The normal distribution follows the standard symmetrical bell-shaped curve. A skewed distribution shows a large distribution of scores at one of the ends of the curve rather than in the center. A bimodal distribution is indicated by large numbers of scores at two different points on the curve, such as at either end.

4. Just because academic achievement test scores and rates of behavioral problems are correlated does not imply that one causes the other. There could very well be a third variable involved that is not apparent that is causing one or the other, or the causes of each of the two variables could be multifaceted.

5. The use of confidence intervals is important in interpreting test scores because it provides you with a band of accuracy or a degree of certainty about how confident you should be that the obtained score is a good measure. The higher the confidence interval, the more assured you can be about the obtained score being a good estimate of the true score.

## APPLICATION ACTIVITY ANSWERS

1. Mean: 90.60     Median: 91     Mode: 91

2. Range: 72 - 112

3. This distribution would be negatively skewed.

4. This group of students scored below the average of what would typically be expected on the WISC-III, by about 2/3 of a standard deviation.

# CHAPTER 6
# WHY WORRY ABOUT RELIABILITY?
# RELIABLE MEASURES YIELD TRUSTWORTHY SCORES

## CHAPTER SUMMARY

In the first few chapters of your text, the concept of test reliability was introduced. Chapter 6 takes you beyond the basics of understanding what reliability is, providing you with a wealth of detail on how test reliability affects educational decision making and what factors affect test reliability. After carefully studying Chapter 6, you will realize the concept of reliability is much more important and more complex than simply how consistently a test measures something.

The chapter begins with a brief overview of how the concept of reliability pertains to educational measures, including a discussion of true scores and types of measurement error. Next, you are introduced to alternate ways of estimating test reliability, including test-retest, parallel form, and internal consistency reliability methods. Following this section, Chapter 6 includes discussions of the use and interpretation of reliability measures, increasing the reliability of tests, and how to understand standard error of measurement. The chapter ends with a discussion of estimating reliability on criterion-referenced measures and the consequences of test reliability on educational classification decisions.

You may think at first glance that the information on test reliability that is presented in Chapter 6 is more detailed and complex than a classroom teacher might need to know. It's true that most classroom teachers will seldom have the need to perform complicated mathematical procedures for determining test reliability, but they will also be greatly empowered by having the basic knowledge of test reliability to help them make informed choices and increase the usefulness of their evaluation procedures. By developing a good basic working knowledge of the key terms and concepts in Chapter 6, you will put yourself in a position of having the skills you need to be a wise consumer of tests and a good producer of your own reliable evaluation materials.

## CHAPTER OUTLINE

Reliability of Educational Measures
    Reliability, True Scores, and Measurement Error
Alternative Ways to Estimate Reliability
    Procedures for Calculating Various Reliability Estimates
Test-Retest (Same Form) Method
Parallel Form Method
Internal Consistency Reliability Estimates
    Split-Half Method for Estimating Reliability
    Kuder-Richardson Reliability Estimates
    Cronbach's Alpha Method for Estimating Reliability

Comparison of Methods for Estimating Reliability: A Summary
Use and Interpretation of Reliability Estimates
How to Increase the Reliability of a Test
Reliability and the Standard Error of Measurement
The Reliability of Criterion-referenced Measures

## GUIDED STUDY EXERCISES

### Definition Exercises

Locate the following terms in your text and briefly define each in the context of educational measurement and evaluation.

Measurement Error:


True Score:


Obtained Score:


Test-Retest Reliability:


Parallel Form Reliability:


Internal Consistency Reliability:


Random Error:


Constant Error:

## Multiple-Choice Questions

Circle the letter of the item that best answers each question.

1. Arla receives a score of 97 on an IQ test. This score is Arla's

   a. true score.
   b. error score.
   c. obtained score.
   d. percentile score.

2. A team of high school math teachers have developed a test for pre-algebra skills. They are interested in having the test adopted for use across the district, but first need to test its psychometric properties. To ascertain the reliability of the test, they each administer it to their students on January 25 and then administer it to the same students on February 25. The scores from the two administrations are correlated. What type of reliability are they measuring?

   a. Internal consistency
   b. Split-half
   c. Parallel form
   d. Test-retest

3. Which of the following is not useful in increasing test reliability?

   a. Increase the reliability of the scoring.
   b. Reduce the variability in test scores.
   c. Make sure the difficulty level is appropriate.
   d. Increase the number of test items.

4. Which form of test reliability is measured by the split-half, Kuder-Richardson, and Cronbach's Alpha procedures?

   a. Parallel form
   b. Test-retest
   c. Internal consistency
   d. Temporal

5. If the proportion of agreement on classification decisions based on test scores is .98, which of the following choices should be assumed?

   a. Classification decisions have been very consistent.
   b. Tests used have perfect reliability.
   c. Reliability of the tests used is also .98.
   d. Classification decisions have been inconsistent.

6. Which of the following reliability methods tends to most underestimate the actual reliability of a measure?

   a. Split-half
   b. Kuder-Richardson
   c. Test-retest
   d. Parallel form method without an intervening time interval

7. For making decisions about students that have important and lasting consequences, what level of reliability coefficient should the tests used in the decision-making process have?

   a. at least .98
   b. at least .96
   c. at least .93
   d. at least .90

8. What level of test reliability coefficients are acceptable for making decisions about groups of students?

   a. As low as .70
   b. As low as .60
   c. As low as .50
   d. As low as .40

## Short-Answer Questions

Provide a brief answer for each of the following questions in the space provided.

1. Describe the relationship between the following: true score, obtained score, and error score.

2. Differentiate between random error and constant error.

3. Why is determining reliability of CRTs more difficult than determining reliability of NRTs?

4. How can the proportion of agreement in classification decisions be artificially inflated?

5. Describe how poor reliability of a test can lead to misclassification of students.

## APPLICATION ACTIVITIES

### Making Tests More Reliable

Your study of Chapter 6 has introduced you to the basics of test reliability in considerable detail and has provided you with several examples of things that threaten the reliability of tests and what can be done to improve it. This activity will help you to synthesize this knowledge by having you make decisions about "real-life" examples of tests that could be used in typical classrooms.

Each of the following four examples of a testing situation describes a test where there is a reliability problem. Your task is to analyze each of the situations based on the information that is provided and decide what might be done to improve test reliability. Compare your responses with the answers at the end of this chapter.

1. Candice, a middle school science teacher, has developed an end-of-unit exam to give to each of her five general science classes. She has put considerable effort into developing a challenging and comprehensive test and has used it at the end of each semester for two years in a row. In analyzing the four sets of test scores, she notices a pattern of inconsistency, leading her to believe that the test has a reliability problem. The test is composed of 50 multiple-choice questions. The average grade in the classes has been C- to D+, with many students receiving D and F grades, a few receiving B grades, and only a small handful receiving A grades.

2. Jarrol teaches third grade. He has developed a basic spelling and written language test to use at the end of the year to determine mastery level of the basic academic goals he has been working on in this area. The test is 10 questions long, a combination of fill-in and multiple-choice formats. Since he is also taking a course in measurement and evaluation at the local university, as part of his master's degree requirements, Jarrol decides to calculate the internal consistency reliability of the test for a class assignment. He is dismayed to find that the reliability is quite poor.

3. Cecilia is a high school math teacher who has developed a test to determine if students who have finished the pre-algebra course are ready to move into general algebra, which is a rigorous precollege course. The test is 45 items long, a combination of multiple-choice and completion formats. After using it for a year, she notices several things. Students tend to do very well on the test (average score is 92 percent), but the scores do not seem to predict future performance in the algebra class, and do not correlate well with pre-algebra grades. What kind of reliability problem might be present?

4. Lupe teaches English and writing skills at a high school. He has been given the assignment to develop a test that might be used in the future as a minimum competency test for written language skills needed to graduate from the high school. The test consists of six essay questions designed to elicit students' written language skills in an open-ended manner. Teachers are given some general guidelines and a point system to use in scoring the essays. After a few trial runs with the test, the district measurement specialist is asked to evaluate the test. She finds that the inter-rater reliability of scores is fairly low, about .43. What can be done to improve it?

## ANSWERS TO GUIDED STUDY EXERCISES

### Definition Exercises

*Measurement Error*: The amount of error in a test score; anything that causes a student to score higher or lower than his or her actual ability.

*True Score*: The hypothetical mean of an infinite number of times a student would take the same test under identical testing conditions.

*Obtained Score*: The actual score received when a student takes a test, which includes the student's true score and measurement error.

*Test-Retest Reliability*: A form of test reliability determined by administering a test to the same group of students at different points in time and obtaining a correlation of the two sets of scores.

*Parallel Form Reliability*: A form of test reliability determined by obtaining correlations between two different but equivalent forms of a test.

*Internal Consistency Reliability*: Reliability estimates (including split-half, Kuder-Richardson, and Cronbach's formulas) that are based on how homogenous a test is or how well it correlates with itself.

*Random Error*: Test errors that are unsystematic and undependable in that they differ from one measurement to another.

*Constant Error*: Tests errors that are uniform and systematic or, in other words, would be the same for all cases.

## Multiple-Choice Questions

1. C
2. D
3. B
4. C
5. A
6. B
7. D
8. C

## Short-Answer Questions: Key Points

1. Each test score contains an obtained score, an error score, and a true score. The obtained score is the actual score that is received or derived. The error score is an error component that reflects random variations in scores. The true score reflects the student's actual ability or knowledge. A formula for these three components is as follows: True score plus error score equals obtained score.

2. Random error occurs unsystematically from one measurement to another--the size and direction of the error can vary appreciably from student to student. Constant error occurs in a systematic manner--the size and direction of the error is the same for each student.

3. The main reason that determining reliability of CRTs is more difficult than for NRTs is that CRTs are not designed to emphasize differences among individuals, while NRTs are. Because increasing the variability among test scores of a group of students leads to increased reliability, NRTs tend to have an advantage in this area.

4. The proportion of agreement in classification decisions can be artificially inflated by chance; in some classification decisions, there can be agreement by two or more individuals simply due to coincidence.

5. Poor reliability of a test can lead to misclassification of students because with unreliability comes increased measurement error. The opportunities for both systematic error and chance agreement increase as test reliability decreases.

# APPLICATION ACTIVITY ANSWERS

1. Candice's test is too difficult. With most of the students scoring C- or lower, and only a few scoring higher, she should decrease the overall difficulty of the test while still retaining challenging items. This would increase the variability of group scores.

2. This test is too short. In most cases, 10 items is short enough to reduce reliability estimates. The teacher should try to increase the length of the test, perhaps to 20 questions.

3. This test is too easy. Since the average score is 92 percent, and it correlates poorly with both grades received and future performance, the way to increase reliability would be to make it somewhat more challenging, which would increase the variability of group performance.

4. Since the interrater reliability is too low, a detailed set of scoring criteria should be developed, with specific guidelines for teachers to follow in scoring the essays. This would tend to increase reliability of scoring.

# CHAPTER 7
# WHY WORRY ABOUT VALIDITY?
# VALID MEASURES PERMIT ACCURATE CONCLUSIONS

## CHAPTER SUMMARY

Chapter 7 of your text refers to validity as the cornerstone of educational measurement. Test validity, which is defined as how well a test measures what it is supposed to measure, is the critical ingredient that test developers must establish and test users must look for. This chapter will help you understand the many facets of test validity and how they work together in establishing whether or not a test measures what it is purported to measure.

Chapter 7 begins with a general discussion of validity as the cornerstone of educational measurement and provides a detailed model for understanding the concept of validity. After this introduction, several different approaches to establishing test validity are detailed, including content, criterion-related, predictive, concurrent, construct, and face validity. Ending this section of the chapter is a discussion of how logical and empirical approaches are used to establish validity. The chapter ends with sections that tie the concept of reliability to test validity and help you to understand the various factors that combine to make a test *useful*.

While studying Chapter 7 and completing the exercises in this workbook chapter, try to keep in mind the idea that test validity is the single most important technical aspect of any educational measure. By carefully studying this material, you will develop the basic knowledge and skills that can be used to both develop and select educational measures that will accomplish their tasks.

## CHAPTER OUTLINE

Validity: The Cornerstone of Good Measurements
    An Integrated Concept of Validity
Approaches to Establishing Validity
    Content Validation
    Criterion-related Validation
    Construct Validation
    Validity Depends on Both Logical and Empirical Approaches
    Face Validity
    Factors That Can Reduce Validity
    Interpreting--and Improving--Validity Coefficients
Validity of Criterion-referenced Measures
The Role of Reliability and Validity in Determining a Measure's Usefulness

# GUIDED STUDY EXERCISES

## Definition Exercises

Locate the following terms in your text and briefly define each in the context of educational measurement and evaluation.

Face Validity:


Content Validity:


Criterion-related Validity:


Predictive Validity:


Construct Validity:


Test Usefulness:

**Multiple-Choice Questions**

Circle the letter of the item that best answers each question.

1. The developers of the Academic Aptitude Test demonstrate that their test correlates very highly with the SAT test (coefficient of .87). Which of the following forms of validity does the high correlation suggest?

    a. Concurrent validity
    b. Content validity
    c. Criterion-related validity
    d. Face validity

2. In comparison with NRTs, determining reliability and validity for CRTs is

    a. easier.
    b. more difficult.
    c. less important.
    d. more critical.

3. Which of the following choices must be in place in order for a test to have validity?

    a. The test must have a moderate standard error of measurement.
    b. It must contain empirically derived content.
    c. It should be of the norm-referenced variety.
    d. The test must be reliable.

4. In order for a test to be useful, which of the following characteristics must it possess?

    a. Relevance
    b. Reliability
    c. Usability
    d. All of the above
    e. None of the above

5. Carlos, a college professor, is developing a test to measure knowledge of basic statistics. In writing the test, he reviews a large number of basic statistics textbooks and syllabi from a number of different basic statistics courses. He is conducting the review because he wants to ensure that he is covering the entire area of basic statistics adequately. What type of validity is Carlos seeking to demonstrate?

   a. Criterion-related validity
   b. Face validity
   c. Concurrent validity
   d. Content validity

6. Which of the following statements is true?

   a. A test must be valid before it is reliable.
   b. A test must be reliable in order to be valid.
   c. A test with high reliability will also have good validity.
   d. A test must have a standard error of measurement between 1.0 and 2.0 to be valid.

7. Which of the following statements is false?

   a. Face validity ensures that a test also has construct validity.
   b. Reliability is a necessary precursor to validity.
   c. Content validity ensures that the test items cover the domain being measured.
   d. Determining test validity is more difficult on CRTs than on NRTs.

8. A test publisher claims that its new academic achievement test is valid because it correlates to a high degree with grades students receive in the areas of academic achievement covered by the test. That is, students who do well on a certain area on the test tend to do well in a subsequent course in that area, and students who do poorly on the test tend to receive poor grades in subsequent classes. What type of validity is being referred to?

   a. Predictive validity
   b. Construct validity
   c. Concurrent validity
   d. Face validity

## Short-Answer Questions

Provide a brief answer for each of the following questions in the space provided.

1. Why is reliability considered to be a precursor to validity?

2. List and describe the two components of test validity.

3. Compare and contrast the procedures and problems of determining validity in CRTs versus NRTs.

4. What is the difference between a logical approach to test validation and an empirical approach?

5. Why is the reliability and validity of a criterion measure important when one is determining criterion-related validity?

## APPLICATION ACTIVITIES

### What Kind of Validity Is It?

Chapter 7 and the exercises in this workbook chapter have taken you on an excursion into the concept of validity that has included an overview of several different types of validity. While each form of validity is important in establishing the overall validity of a test, it may be difficult for you at this point to tell them apart. This application activity is designed to help you integrate and solidify your understanding of different test validation procedures.

Carefully read each of the following scenarios of test validation processes, and then decide which form of test validity each is trying to demonstrate. Some scenarios may be utilizing more than one form of test validation, so keep in mind to look for the *big picture*. You can compare your answers with the information at the end of this workbook chapter.

1. A large manufacturing firm employs many persons on assembly lines where they are required to put together complicated electronic components within a short period of time. The management of the company is frustrated with their inability to determine beforehand which potential employees have a knack for this kind of work and which employees will continually struggle with the task. The company management hires an educational measurement specialist to work with a production manager in developing a test that will allow them to determine which potential employees are likely to be successful on the assembly line. The measurement specialist conducts research to determine how well scores on the test correlate with later productivity and quality on the assembly line. What type of validity was being sought with this research?

2. The developers of the Behavior Disorders Rating Scale have conducted a study in which they had the scale completed on a group of students, and at the same time, had the same students rated with a similar scale, the Behavior Problem Checklist. Following the data collection, a correlation coefficient between the two rating scales was obtained. What type of validity might be demonstrated by this?

3. The developers of a new intelligence test are trying to demonstrate that their new instrument is indeed measuring "intelligence." They conduct a study wherein scores from the new test are compared with a number of other measures and behaviors that are believed to measure intelligence. The purpose of the study is to prove that the test is actually a valid measure of intelligence. What type of validity are they trying to demonstrate?

4. Andrea, a high school science teacher, has compiled a test for measuring general science skills that students ought to have when they leave high school. She asks several of her colleagues to look it over to see if it "looks like" it is really measuring science knowledge and not including knowledge from related fields like math. Her colleagues agree that it does indeed appear to be measuring science knowledge. What type of validity has been demonstrated?


## ANSWERS TO GUIDED STUDY EXERCISES

### Definition Exercises

*Face Validity*: The degree to which a measurement instrument appears to measure what it is intended to measure.

*Content Validity*: The extent to which the content of a test's items represents the entire body of content to be measured.

*Criterion-related Validity*: The extent to which one can infer from an individual's score on a test how well he or she will perform on some other valid external criterion.

*Predictive Validity*: How well a test measures, predicts, or estimates future performance on some valued criterion other than the test itself.

*Construct Validity*: The extent to which a test measures the particular hypothetical construct or trait that it is intended to measure.

*Test Usefulness*: In order for a test to be useful, it must have good validity (relevance plus reliability) and good usability (be easy to use, not be overly technical or time consuming).

### Multiple-Choice Questions

1. A
2. B
3. D
4. D
5. D
6. B
7. A
8. A

### Short-Answer Questions: Key Points

1. Reliability is a precursor to validity for a very simple reason--a test cannot effectively measure what it is supposed to measure if it is not consistent. The more error introduced into a test through poor reliability, the lower the probability that the test is measuring what it was designed to measure.

2. For a test to be valid, it must be both reliable and usable. If a test is reliable, it will provide a consistent measure. If a test is usable, it will be affordable, of reasonable length, and reasonably easy to administer and interpret.

3. Validity approaches that depend on the use of correlation coefficients may be less useful for criterion-referenced tests than they are for norm-referenced tests, due to the limitation variation in criterion-referenced scores. Evidence of content validity for criterion-referenced instruments is particularly important, while obtaining evidence of construct and criterion-related validity is particularly important for norm-referenced tests.

4. Logical approaches to validity involve a rational process of establishing content validity, such as obtaining good "face" validity or establishing the appearance of validity. Empirical approaches to test validation, on the other hand, involve statistical methods of calculating predictive and criterion-related forms of validity.

5. When attempting to establish criterion-related validity, it is imperative that the criterion measure used have acceptable levels of both reliability and validity. If the criterion measure does not have adequate reliability, the increased measurement error may result in the appearance of a strong correlation between the two tests, but this relationship could be based on chance. If the criterion measure is not valid, there is no point in using it to establish validity of another instrument.

# APPLICATION ACTIVITY ANSWERS

1. Predictive validity

2. Concurrent validity

3. Construct validity

4. Face validity

# CHAPTER 8
# CUTTING DOWN TEST SCORE POLLUTION:
# THE INFLUENCE OF EXTRANEOUS FACTORS

## CHAPTER SUMMARY

By this point in your academic career, you have no doubt taken hundreds of tests. Most of these tests have probably been academic achievement tests used to measure your progress in school, and perhaps to determine your grades. You have probably taken a few tests of other types, such as intelligence or aptitude tests, career interest inventories, and perhaps some tests of social or emotional adjustment. With all of your testing experiences at this point, it would be a safe guess to say that the results of some (if not many) of the tests you have taken have been influenced by extraneous factors, such as your approach to test taking, the way the instructions were given by the examiner, or how well or poorly constructed the test items were. Chapter 8 in your text deals with these very issues--how various influences we refer to as "extraneous" can have the unintended effect of influencing test scores and even threatening the validity of tests.

The chapter begins with a discussion of "test score pollution" and then gives you a tour through different types of extraneous factors that can influence performance on cognitive tests. Some of these factors include the test-taking skills and typical ways of responding that a subject brings to the test, the effects of personal factors such as anxiety and motivation, how the test is administered, the effects of coaching and practice, and test bias. A similar type of overview is provided for extraneous factors that can influence affective measures (measures of social-emotional functioning). Some of the sources of test pollution that can influence affective measures include deliberate attempts to incorrectly answer questions, problems of omitting and misinterpreting test items, examinees' lack of insight into their motivations or fooling themselves, and answering questions in a random fashion. The chapter ends with a discussion of how to reduce or eliminate various extraneous factors from tests and provides two sets of guidelines for you to follow in doing this.

When you read the text chapter and go through the related exercises and activities in this workbook, try to think of examples from your own experience as a test taker, or potential situations from testing in your specialty area, that relate to the concepts being stressed. That way, these everyday problems with testing will seem more relevant to you, and you will begin to understand the caution and precision that you will need to exercise as a developer and administrator of tests. While these so-called extraneous influences are numerous, there are some specific steps you can take that will minimize their effects or eliminate them altogether.

## CHAPTER OUTLINE

A Taste of "Test Score Pollution"
Extraneous Factors That Can Influence Performance on Cognitive Tests

Test-Taking Skills
  Testwiseness
  Response Sets
  Anxiety and Motivation
  Administrative Factors
  Coaching and Practice
  Test Bias
Extraneous Factors That Can Influence the Results of Affective Measures
  Social Desirability
  Faking
  Item Omission
  Problems of Interpretation
  Self-Deception and Lack of Insight
  The Acquiescence Response Set
General Strategies for Eliminating or Reducing the Impact of Extraneous Factors

## GUIDED STUDY EXERCISES

### Definition Exercises

Locate the following terms in your text and briefly define each in the context of educational measurement and evaluation.

Extraneous Variables:


Testwiseness:


Response Set:


The Acquiescence Set:


Motivation:

Examiner Effects:


Test Bias:


Social Desirability:


Faking:



**Multiple-Choice Questions**

Circle the letter of the item that best answers each question.

1. One extraneous factor that may influence the results of tests where students are asked to provide the "best" answer is that

    a. the "best" answer is often incorrect.
    b. a student may understand the concept but still give an incorrect answer.
    c. the more a student knows about a subject, the more difficult the choice becomes.
    d. this questioning method has been shown to lead to extremely poor test reliability.

2. Which of the following statements regarding "guessing" on a test is false?

    a. Appropriate guessing is a valuable test-taking skill.
    b. Some students hesitate to guess on tests because they are discouraged from doing it during regular instruction.
    c. Research cited in the text indicates that 30 to 40 percent of test takers guess at answers.
    d. A students "true" ability is more readily indicated when he or she systematically eliminates options and makes appropriate guesses.

3. Absolute terms such as *never* and *always* in test answer options

   a. may clue "testwise" students to avoid those items.
   b. are recommended as a good way of developing "distractor" items.
   c. are a tip-off that the answer option is probably correct.
   d. are *never* to be used in constructing test items.

4. Which of the following statements concerning motivation and anxiety is true?

   a. Students who become *overly* motivated may become anxious.
   b. Large amounts of anxiety tend to increase performance on tests.
   c. Highly motivated students tend not to perform well on tests.
   d. There is a large, negative correlation between anxiety and performance on cognitive tests.

5. On true-false tests, it has been shown that respondents are more likely to select the option designated as "true" when they are uncertain. How has this phenomenon been labeled?

   a. The acquiescence set
   b. The option length set
   c. The positional preference set
   d. The social desirability set

6. Which of the following situations best characterizes the existence of test bias?

   a. The groups receiving lower scores are known to be equally able.
   b. The groups scoring lower are females or minorities.
   c. The test is an ability test.
   d. The standardization group is composed of majority group students.

7. On a personality test that asked respondents to answer either true or false to the statement "I believe that criminals can change their ways with the right kind of help," Diane responded "true" even though she really believed the statement is false. What is this scenario an example of?

   a. Self-deception response set
   b. Social desirability response set
   c. Acquiescence response set
   d. Problems of interpretation

8. The most frequent approach to reducing "faking" on locally developed affective measures is to

   a. develop a measure that is too sophisticated to fake answers on.
   b. tell examinees that negative consequences will come to them if they fake.
   c. convince examinees that a truthful response is important.
   d. develop a sophisticated "lie" scale such as the one used on the MMPI.

## Short-Answer Questions

Provide a brief answer for each of the following questions in the space provided.

1. How is a social desirability response set different from faking on affective tests?

2. Summarize what we know about the effects of coaching and practice on taking tests.

3. How does familiarity with the examiner tend to affect a student's performance on a test?

4. How do motivation and anxiety level impact test performance?

5. With respect to using machine-scored forms (where the students fill in the bubble) for tests, can this impact test performance, and what has been suggested to be the minimum grade level for students to be at before they use them with tests?

## APPLICATION ACTIVITIES

### Identifying the Sources of Test Pollution

David is about to administer an end-of-unit test that he has developed for his 11th grade U.S. History classes, which number three. Unfortunately, it has been a while since David has studied educational testing and measurement, and he has gotten sloppy in his testing practices. You guessed it--David is about to introduce "test pollution" into his methods of evaluating his students.

Carefully observe how the following scenario unfolds, and make a list of each thing David does that might introduce extraneous variability into the test scores. Then categorize each of the practices by the category or categories they are listed under in the text chapter. You can check your responses with the information provided at the end of this workbook chapter.

*David has developed a 45-item test for his U.S. History students. The first part of the test is a true-false section that has 20 items in it, 15 of which have "true" as the correct answer and 5 of which have "false" as the correct answer. Another section of the test has 15 multiple-choice questions in it. Of these 15 multiple-choice questions, the correct response choice is also the longest choice for 10 items. The remaining 10 test items are short-answer or matching items. David administered this test to his three U.S. History classes on a Friday (first, third, and sixth periods). Even though each class period is 50 minutes long, David imposed a strict time limit of 35 minutes for the test. He instructed his students to "go with your first impression of what the correct answer is--it is best not to change your answers." He did not give instructions in the same way to the three classes; he was more direct and specific during the first and third periods, but rather vague during the sixth period. By the time his students have all taken the test and he has had a chance to score them all, some students have done better than they really deserved to, and others have not done as well as they were capable of doing.*

## ANSWERS TO GUIDED STUDY EXERCISES

### Definition Exercises

*Extraneous Variables*: Characteristics, occurrences, or behavior patterns that can result in an apparently valid test becoming an inaccurate measure of what it was designed to measure.

*Testwiseness*: The ability to use "clues" in the test to obtain a higher score than is deserved.

*Response Set*: A tendency to respond in consistent ways on tests that may be related to the types of options chosen, the speed with which work is done, or the probability of guessing when the correct answer is not known.

*The Acquiescence Set*: A type of response set whereby respondents are more likely to select the option that is designated as "true" when they are uncertain.

*Motivation*: The students' desire to perform on a test as best they can.

*Examiner Effects*: Characteristics or behaviors of the test examiner (such as familiarity or rapport with the examinee) that can influence the performance of the person(s) taking the test.

*Test Bias*: A condition where individuals from different groups, although equally able, do not have equal probabilities of success on a test.

*Social Desirability*: A response set where examinees will respond to items in a way they believe to be most socially appropriate, regardless of their true feelings.

*Faking*: A condition where a subject taking an affective test attempts to manipulate or "fake" responses in order to achieve a desired result.

## Multiple-Choice Questions

1. B    5. A
2. C    6. A
3. A    7. B
4. A    8. C

## Short-Answer Questions: Key Points

1. On affective tests, a social desirability response set involves the examinee being influenced through subtle social pressure to select the "right" choice. Faking, on the other hand, involves a deliberate attempt on the part of the examinee to choose a response that is other than the way he or she truly feels in order to achieve desired results.

2. Coaching or practice on standardized tests may have significant effects for some students, but for most students the effects are likely to be modest. It is generally thought that the time spent on practice or receiving coaching for standardized tests is no more productive (and more costly) than time spent in school.

3. There is some evidence that scores on individually administered cognitive tests may be maximized if the examinee is familiar and comfortable with the examiner. It is likewise true that for some students, unfamiliarity and discomfort with the examiner may lead to lowered test scores.

4. Students who are highly motivated tend to perform substantially higher on standardized tests than do poorly motivated students. Students who become overly motivated may become anxious, and high levels of anxiety are related to poorer test performance.

5. When using machine-scored "fill in the bubble" forms, students may miss items they would otherwise answer correctly, if the answer format is confusing or distracting. It is suggested that separate answer sheets not be used with students below the fourth grade level.

# APPLICATION ACTIVITY ANSWERS

## Which of David's Practices Introduced "Test Pollution"?

| PRACTICE THAT INTRODUCED "TEST POLLUTION" | CLASSIFICATION OF THAT PRACTICE |
|---|---|
| Out of 15 multiple-choice items, the correct answer was also the longest answer on 10 of them | *Response Sets Related to Option Length.* There is a tendency for test makers to make the correct choice the longest, and there is also a tendency for some test takers to choose the longest option when they do not know the answer. |
| Out of 20 true-false items, 15 had "true" as the correct response | *Response Sets Related to Item Construction*: Respondents are more likely to select the "true" option when they don't know the answer. |
| Instructions were given vaguely | *Administrative Factors/Examiner Effects*: Examiner should provide clear and consistent instructions. |
| Examiner imposed time limit on the test | *Test-taking Skills/Response Sets*: Some students cannot perform well under time limits, and some students will have a "speed versus accuracy" response set. |
| Examiner told students not to guess | *Test-taking Skills*: Appropriate guessing is a valuable skill and is more likely to provide a measure of true ability. |

# CHAPTER 9
# CONSTRUCTING YOUR OWN ACHIEVEMENT TESTS -- DECIDING WHEN AND HOW TO DO SO

## CHAPTER SUMMARY

You have learned by now that the type of tests most commonly administered in schools are teacher-made tests. Chapters 9, 10, and 11 in your text focus on the process of constructing your own achievement tests, laying out a logical order for developing a plan or blueprint for your tests, actually developing test items, and then finishing the test, administering it, and using the data in a professional manner. Chapter 9 is the first part of this sequence. It lays a foundation for understanding the basics of developing your own tests.

The chapter begins with a discussion of why teachers might want to construct their own achievement tests and helps you to consider the many advantages that teacher-made tests can have over commercially published standardized tests. The next section of Chapter 9 deals with a step in the testing process that is somewhere between using commercially published tests and developing your own. It involves adapting commercially available tests for local needs. Specifically, you are introduced to techniques of customizing standardized test scoring and using item banks to make these tests more appropriate to your specific needs. The majority of Chapter 9 deals directly with the process of constructing your own achievement tests. The specific focus in this chapter is on the *planning* aspect of the process. Chapter 9 concludes with a detailed description of three steps in the planning process: (1) clarifying educational objectives, (2) specifying what you want to test, (3) developing a test blueprint.

Consider chapters 9, 10, and 11 from your text to be partners in helping you develop the skills you need to create and use your own educational tests. If you become a classroom teacher, it is likely that the majority of tests you use will be tests that you create. And yet, few teachers have undergone the training and practice necessary to maximize the effectiveness of the test-building activities. By using these three chapters carefully, you will begin to acquire the skills you need to be an "expert" test creator.

## CHAPTER OUTLINE

Why Construct Your Own Achievement Tests?
Adapting Commercially Available Tests for Local Needs
Constructing and Using Local Achievement Tests
Steps in Planning Achievement Tests
Clarifying Instructional Objectives
Specifying What You Want to Test
Developing a Test Blueprint

# GUIDED STUDY EXERCISES

## Definition Exercises

Locate the following terms in your text and briefly define each in the context of educational measurement and evaluation.

Test Blueprint:


Pre-instructional Assessments:


Interim Mastery Tests:


Mastery Tests:


Educational Taxonomy:


Instructional Objectives:


Item Banks:

**Multiple-Choice Questions**

Circle the letter of the item that best answers each question.

1. Which of the following statements best indicates the type of instruction or practice in test construction that most teachers receive?

    a. Few teachers receive sufficient instruction or practice in item writing or test construction.
    b. Most teachers receive sufficient instruction or practice in item writing or test construction.
    c. All teachers receive sufficient instruction or practice in item writing or test construction.
    d. No teachers receive sufficient instruction or practice in item writing or test construction.

2. Which of the following is *not* a potential advantage of locally developed achievement tests over commercially published standardized achievement tests?

    a. They tend to be more relevant to local needs and situations.
    b. They can be administered more frequently.
    c. They tend to have higher levels of test reliability.
    d. They can provide more detailed information on specific objectives.

3. Which of the following statements represents the opinion of the authors of your text about including some material on tests that has *not* been taught?

    a. Including questions on material that has not been taught is always unethical.
    b. It is useful to know how students do on related material that has not been taught.
    c. It is a waste of time to test on related material that has not been taught.
    d. The practice of testing on related material that has not been taught increases test validity.

4. In a survey of high school teachers, Salmon-Cox (1981) identified factors that teachers use in making evaluative decisions about students' achievement. Which of the following factors was relied upon most heavily?

    a. Classroom interactions with students
    b. Homework assignments
    c. Standardized achievement test scores
    d. Teacher-made tests

5. Which of the following statements about the three stages of the testing process (planning, development, and application) is correct?

    a. The three stages are all interrelated.
    b. Application exists independently of planning and development.
    c. Planning exists independently of application and development.
    d. The three stages all exist independently of each other.

6. From least common to most common, which is the correct order of the three most common parts of Bloom's taxonomy of educational objectives?

   a. Knowledge, application, comprehension
   b. Application, knowledge, comprehension
   c. Application, comprehension, knowledge
   d. Knowledge, comprehension, application

7. One of the most frequent criticisms of educational tests stems from the fact that they have been found to focus primarily on

   a. analysis items.
   b. rote memorization and recall of facts.
   c. evaluation of inappropriate information.
   d. comprehension of information not linked to instructional objectives.

8. Which of the following statements is a good argument for reconsidering instructional objectives periodically?

   a. It is easier to test for how well students have achieved some instructional objectives than others.
   b. Instructional goals seldom change over time.
   c. Most instructional objectives are not utilized in assessment.
   d. There are legal constraints requiring teachers to periodically change instructional objectives.

## Short-Answer Questions

Provide a brief answer for each of the following questions in the space provided.

1. Why is it important to clarify instructional objectives before developing a test?

2. List four different situations in which developing your own achievement test would be preferable to using a commercially published standardized achievement test.

3. List two methods of adapting commercially published standardized achievement tests in order to make them more relevant for local needs.

4. What are some of the potential problems that may occur when testing is not appropriately linked to teaching?

5. List the different components of Bloom's taxonomy of educational objectives for the cognitive domain.

## APPLICATION ACTIVITIES

### Practice in Using a Taxonomy of Educational Objectives

One of the points that is frequently emphasized in Chapter 9 of your text is that clarifying and specifying instructional objectives is an important part of the process of developing good test items on achievement tests. This application activity will provide you with experience in using a taxonomy of educational objectives to specify what you want to test.

The first step in the activity is to select an instructional area and specific topic from your specialty area that you want to work with. If you do not have an educational specialty area at this point, simply select an area that interests you (for example, basic reading skills at the primary level, pre-algebra, high school geography, etc.). After you have selected your area and specific topic, develop a list of educational objectives for it by using the six areas from Bloom's taxonomy in the table below this section as a guide. It is advisable to review the section from Chapter 9 on specifying educational objectives (including the information in Table 9.1 and Figure 9.2) before tackling this activity. When you are prepared to begin, fill in the blank sections in the table that correspond to the six different areas of the taxonomy. Since your choice of an instructional area and specific objectives will be unique, there can be no specific answers to compare your work to. However, an example from the area of basic reading and written language is provided at the end of the chapter to use as a guide.

| Instructional Area/Topic Selected: | |
|---|---|
| **Taxonomy Area** | **Educational Objective** |
| 1. Knowledge | |
| 2. Comprehension | |
| 3. Application | |
| 4. Analysis | |
| 5. Synthesis | |
| 6. Evaluation | |

## ANSWERS TO GUIDED STUDY EXERCISES

### Definition Exercises

*Test Blueprint:* An integrated master plan for developing a test based on course content, instructional objectives, and learning activities.

*Pre-instructional Assessments:* Test procedures for ascertaining what skills students have already mastered prior to beginning a new unit of instruction.

*Interim Mastery Tests:* Tests administered before moving from one instructional unit to another, given in order to determine whether or not students have mastered the critical concepts from the previous unit.

*Mastery Tests:* Tests that are administered at prespecified times during the year to obtain information useful in making decisions such as assigning grades, deciding on promotion, and determining eligibility for special programs.

*Educational Taxonomy:* A generic name for educational classification systems (such as Bloom's taxonomy) for categorizing levels of cognitive skill or types of behavior.

*Instructional Objectives:* Lists of specific skills and knowledge that have been determined to be the most important things for students to learn in an instructional unit.

*Item Banks:* Large collections of test items that cover specified instructional or content areas.

## Multiple-Choice Questions

1. A
2. C
3. B
4. D
5. A
6. C
7. B
8. A

## Short-Answer Questions: Key Points

1. Clarifying instructional objectives should be done before developing a test to make sure that the content of the test is consistent with the content of the class and to determine what are the most important educational objectives to be tested.

2. Some situations where it would be desirable to develop a teacher-made achievement test rather than use a commercially published standardized test include:

   - The teacher desires to tailor the test items to specific instructional objectives.
   - There is a need to test frequently rather than at long time intervals.
   - There is a need to obtain the test results quickly.
   - Identification of specific learner needs is an important objective.
   - There is a need to make the test consistent with district or state curricula.
   - It is important to cover a specific area of content in considerable detail.

3. Some methods of adapting commercially available tests for local needs include customized test scoring and using previously developed item banks.

4. When testing is not appropriately linked to teaching, the test results will not provide enough information on student mastery of the curriculum and will be of little or no benefit in developing remedial education plans.

5. Bloom's taxonomy of educational objectives for the cognitive domain includes the following components: knowledge, comprehension, application, analysis, synthesis, and evaluation.

# APPLICATION ACTIVITY ANSWERS

| Instructional Area/Topic Selected: BASIC READING AND WRITTEN LANGUAGE SKILLS ||
|---|---|
| **Taxonomy Area** | **Educational Objective** |
| 1. Knowledge | Students will decode words from the basic 100-word reading list at 80 percent accuracy. |
| 2. Comprehension | Given a short sentence, students will be able to paraphrase the original sentence and rewrite in other words. |
| 3. Application | Students will correctly write their home address and the names of persons in their families. |
| 4. Analysis | Given pairs of CVC words and CVC words with "e" endings (e.g., *hop* and *hope*), students will correctly sight read each part of the pair. |
| 5. Synthesis | Students will write a one-paragraph story about their favorite activity, using words from the basic 100-word reading list. |
| 6. Evaluation | Given three sample sentences containing errors, students will identify error patterns with 60 percent accuracy. |

# CHAPTER 10
# STEPS IN DEVELOPING GOOD TEST ITEMS FOR YOUR ACHIEVEMENT TESTS

## CHAPTER SUMMARY

Now that your study of Chapter 9 has provided you with a solid background on reasons for developing your own achievement tests and some general best practices, Chapter 10 will help you develop the "nuts and bolts" skills you need to actually develop and write good quality test items. At this point in your academic career, you have undoubtedly been exposed to (or perhaps been the victim of) a wide variety of teacher-made tests. Think for a moment about some of the tests you have taken that you considered to be fair, challenging, and well designed. Then compare these tests with other tests you may have taken that you considered to be unfair, poorly designed, and perhaps downright bad examples of what tests should be. What things made the difference between the well designed and poorly designed tests? Chapter 10 will help you acquire the skills you need not only to make better judgments about tests but to write your own test items with a great deal of proficiency.

Chapter 10 begins with an overview and discussion of two main steps in developing achievement tests: selecting an item format and preparing test items. Then you are provided with some general guidelines that apply to several, if not most, test item formats. Once you have this basic foundation of test-writing knowledge, Chapter 10 provides you with some very specific and useful rules and guidelines to follow in writing specific types of test items, including true-false, multiple choice, matching, short answer, and essay format. The chapter ends with a summary of some of the main points of information and an interesting story about how much influence the way tests are written can have on instruction and learning.

As a potential classroom teacher, keep in mind that most of the tests you may eventually administer will probably be tests that you have developed and written yourself. By mastering the information in Chapter 10 and keeping it as a resource for your future use, you will be preparing to write the kind of tests that your own students will remember as "good" examples in the future.

## CHAPTER OUTLINE

Steps in Developing an Achievement Test
Selecting an Item Format
Preparing Test Items
General Guidelines That Apply to Several Item Formats
Guidelines for Writing True-False Items
Guidelines for Writing Multiple-Choice Items
Guidelines for Writing Matching Exercises
Guidelines for Writing Completion or Short-Answer Items

Guidelines for Writing and Scoring Essay Questions
In Summary

## GUIDED STUDY EXERCISES

### Definition Exercises

Locate the following terms in your text and briefly define each in the context of educational measurement and evaluation.

"Objective" Test Items:


"Subjective" Test Items:


True-False Tests:


Multiple-Choice Tests:


Matching Exercises:


Completion Tests:


Essay Tests:


Stem:

Distractors:

## Multiple-Choice Questions

Circle the letter of the item that best answers each question.

1. Which of the following is a general guideline that applies to writing test items in several formats?

    a. "Trick" questions are suitable in any format.
    b. Items should be written as simply as possible.
    c. The author of the item should be the sole source for determining how appropriate the item is.
    d. The format of the test can be an appropriate source of distraction.

2. Using direct quotations from the text in writing multiple-choice test items

    a. is appropriate as long as copyright permission is secured.
    b. has been determined to be an illegal practice.
    c. is usually an inappropriate practice.
    d. is considered to be a good method of linking the curricula and the test.

3. It is recommended that true-false test items

    a. be completely avoided due to the prevalence of technical problems.
    b. be used only when correction formulas are utilized in scoring.
    c. should be used with about the same frequency as multiple-choice items.
    d. should not be the only or predominant means of testing.

4. What is the recommended number of options on multiple-choice questions?

    a. Three to five
    b. Four
    c. Four to six
    d. Two to four

5. On multiple-choice items, incorrect answer options should be written so that they

    a. clearly appear to be incorrect.
    b. seem plausible to students who have not mastered the material.
    c. seem plausible to students who have mastered the material.
    d. could seem correct if certain conditions were in place.

6. Of the following choices, which is the most frequently omitted instruction on matching items?

    a. Which column contains the stem and which column contains the distractors
    b. Which column the choices should be made from
    c. The listing of any time limits that are involved
    d. Whether options can be matched with more than one item

7. Which of the following is a recommended practice for developing matching exercises on tests?

    a. Don't confuse the students by having all the options plausible.
    b. Avoid exercises where all items are used and each option is used once and only once.
    c. Limit the number of items to a maximum of four to five.
    d. Place the two columns on different pages to avoid confusion.

8. Which of the following is *not* a reason for the popularity of short-answer completion questions?

    a. Ease of construction
    b. Ease of scoring
    c. Large numbers of questions in limited space
    d. Higher reliability than true-false or multiple choice

## Short-Answer Questions

Provide a brief answer for each of the following questions in the space provided.

1. Which type of test items can be written in ways that test higher-level cognitive skills as well as recall of knowledge?

2. Develop one example where a test item has a clue that "gives away" the test answer.

3. Why has the use of true-false test items been on the decline?

4. How do "trick" questions differ from good options that seem plausible to students who have not mastered the material?

5. Why should "all of the above" or "none of the above" seldom be used as answer options in multiple-choice questions?

APPLICATION ACTIVITIES

What's Wrong with These Test Items?

One of the key skills you can acquire in learning how to write good test items is to learn how to recognize when a test item is poorly constructed and what you can do about it to improve it. After reviewing the information from Chapter 10 on general and specific rules of item writing, look at the following five examples of items and determine what kind of problem is present in the way the item is written and what could be done to improve it. Check your responses with the information provided at the end of this workbook chapter.

A. Which of the following is a good explanation for the high rates of unemployment, poverty, and crime in urban minority communities?

   a. Differences in cultural practices and values with the dominant culture
   b. Feelings of hopelessness due to continuing lack of economic opportunities
   c. The pervasive influence of gang activities and illegal drugs
   d. The historical impact of racist attitudes in American culture

B. When George Washington and his crew crossed the Delaware River, what kind of material was the hat he was wearing made of?

C. T  F     President John F. Kennedy was shot by Leigh Harvey Oswald in Dallas, Texas.

D. Which city is the capital of the state of Oregon?

   a. Salem
   b. New York
   c. Los Angeles
   d. Mexico City

E. Multiple-choice test items

   a. should be written so that there is only one correct or best answer.
   b. should be written so that the distractors are not plausible.
   c. should be written so that "none of the above" is a frequent choice.
   d. should be written so that the stem should not make sense until the options are read.

## ANSWERS TO GUIDED STUDY EXERCISES

### Definition Exercises

*"Objective" Test Items*: Test items that follow a true-false, multiple-choice, or matching format.

*"Subjective" Test Items*: Test items that are developed in short-answer or essay format.

*True-False Tests*: A test format where the subject must mark whether a statement is "true" or "false."

*Multiple-Choice Tests*: A test format where the subject must choose among a number of potentially correct alternatives that relate to a statement or question.

*Matching Exercises*: A test format where the subject is presented with two columns of information and must select which option in the second column best matches each item in the first column.

*Completion Tests*: A test format that consists of questions that can be answered with a word or short phrase, or a statement having one or more omitted words, which the subject is supposed to complete.

*Essay Tests*: A test format consisting of questions that must be answered in some detail in writing.

*Stem*: A statement or question that serves as the stimulus in a multiple-choice test item.

*Distractors*: The options or alternatives from which the subject must select a plausible answer in a multiple-choice test item.

## Multiple-Choice Questions

1. B     5. B
2. C     6. D
3. D     7. B
4. A     8. D

## Short-Answer Questions: Key Points

1. Virtually every type of test item can be written in ways to test higher-order thinking skills rather than simple recall of information.

2. The following is an example where a test item has a clue that "gives away" the answer:

   *Which of the following are examples of good practices in writing essay examinations?*

   *a. Restricting questions to learning outcomes that are difficult to measure with other formats*
   *b. Providing guidelines as to time limits and amount of information expected*
   *c. Using several narrowly focused items rather than one broad item*
   *d. Developing a scoring system before administering the test*
   *e. All of the above*

   In this case, the plural use of the words *examples* and *practices* tip you off that there is probably more than one correct answer. Since none of the distractors seems obviously wrong, and the only option that includes more than one choice is e, you are led to choose that answer.

3. The declining use of true-false test items is probably due to beliefs that these items test only trivial details, are ambiguous, encourage rote learning, and are susceptible to high scores from guessing. In reality, true-false items can be written in a way so that these criticisms are not true.

4. Good distractor options do not trick students who have mastered the concepts into selecting an incorrect answer. Trick questions tend to confuse trivial detail with important concepts, thus throwing off students who may have reasonably mastered the material.

5. "All of the above" or "none of the above" as answer options in tests should be used sparingly because in most cases they are not plausible distractors and tend to be overused placeholders.

# APPLICATION ACTIVITY ANSWERS

A. This multiple-choice item has several answers that could easily be construed as being correct, depending on the context in which you were reading the item. A good rule to follow is to have only one correct or best answer on objectively scored tests.

B. This item tests for trivial detail rather than important ideas or concepts.

C. This is a "trick" question, as Oswald's first name was spelled "Lee" rather than "Leigh." This is the kind of trick question that creates cynical students. Avoid using trick questions that confuse trivial details and major issues.

D. Unless you are dealing with an extremely naive student group, the distractor items in question D are not plausible enough to be taken seriously. Most mid-elementary students who have studied U.S. geography would probably be able to deduce that A is the correct answer because they can rule out the others. It would be better to include such distractors as "Portland," "Eugene," and "Olympia."

E. This item breaks the rule of not repeating words in each option that could be placed in the stem. It would be better to have the stem read "multiple-choice test items should be written so that" and remove "should be written so that" from the distractors.

# CHAPTER 11
# THE PROCESS OF BECOMING AN EXPERT TESTER: ASSEMBLY, ADMINISTRATION, AND ANALYSIS

## CHAPTER SUMMARY

In your study of measurement and evaluation in the classroom, you have thus far been introduced to a variety of ideas, concepts, and techniques to help you understand the basics of how testing is used in educational settings. Now that you have a basic foundation in educational testing, Chapter 11 introduces you to some very specific methods and techniques to help you become more expert in testing. The goal of this chapter is to provide you with the tools to enable you to go beyond simply understanding and administering tests, to utilizing tests in a manner that helps you to maximize their benefits.

Chapter 11 begins with an overview of the three cyclical components of testing: assembly, administration, and analysis. Then each of these three areas is covered individually and in detail. In addition, sections on the best practices in scoring tests and how to discuss test results with students are provided in order to broaden your understanding of the cyclical process of testing.

Have you ever taken a test where even though you understood the material pretty well, the way the test was written produced confusion and frustration? Some students refer to this type of test as having "trick" questions, but in reality the confusion and frustration most often results from a lack of systematically understanding and applying the three-part cyclical testing process on the part of the test-giver, rather than in deliberate attempts to trick students. Chapter 11 in your text and the exercises and activities in this chapter will help you to gain the tools for making tests more applicable to your instructional goals and making the test-taking experience more positive and helpful for your students.

## CHAPTER OUTLINE

Assembling the Test
    Collecting Test Items
    Reviewing Test Items
    Formatting the Test
    Preparing Directions for the Test
Administering the Test
    Standards for Test Administration
Scoring Tests
Discussing Test Results with Students
Analyzing Test Items
    Overview of Item Analysis Procedures
    A Step-by-Step Analysis Procedure for Classroom Use
    Using Item Analysis for Revision and Improvement

Cautions about Interpreting the Results of Item Analyses
Item Analysis Procedures with Criterion-referenced Tests

## GUIDED STUDY EXERCISES

### Definition Exercises

Locate the following terms in your text and briefly define each in the context of educational measurement and evaluation.

Assembly:

Administration:

Analysis:

Test Item File:

Practice Tests:

Test Security:

Difficulty Level:

Coverage:

Item Analysis:


Indices of Discrimination:



**Multiple-Choice Questions**

Circle the letter of the item that best answers each question.

1. A test item file is easier to create by developing items

    a. at the end of each unit.
    b. on a monthly basis.
    c. on a day-to-day basis.
    d. at the end of each week.

2. When one test item provides an unnecessary clue to the answer of another, the

    a. items are not independent.
    b. test reliability will be lower.
    c. standard error of measurement will be higher.
    d. items are independent.

3. Items with similar formats

    a. should not be used on the same test.
    b. should be spread out over an entire test.
    c. should be grouped together.
    d. will decrease testing time if they are presented apart.

4. The best way to help students understand what they are to do on a test is to

    a. read the directions twice before they start.
    b. administer a practice test that is similar in format and content to the actual test.
    c. ask them to review the materials to be covered prior to the test administration.
    d. give instructions orally rather than in writing.

5. Sixty minutes of testing time without a break is the maximum recommended for students at which of the following grade level ranges?

   a. K-3
   b. 10-12
   c. 7-9
   d. 4-6

6. Which of the following is considered to be the best method for reducing cheating on tests?

   a. Using alternate seating arrangements
   b. Carefully monitoring test administration
   c. Developing parallel test forms
   d. Rewarding or reinforcing students who do not cheat

7. The primary purpose for doing item analyses is to

   a. determine whether or not students are cheating on a test.
   b. increase the test-retest reliability of an exam.
   c. improve the way a specific test item functions.
   d. make decisions on improving the curriculum covered by the test.

8. Which of the following statements regarding differences in the use of item analyses between norm-referenced and criterion-referenced tests is true?

   a. Criterion-referenced tests tend to have higher levels of difficulty if the instruction has been effective.
   b. There are few differences in using item analysis procedures between the two types of tests.
   c. Discriminating between high and low achievers is the goal for both types of tests.
   d. With both types of tests, low indices of discrimination always indicate defective test items.

**Short-Answer Questions**

Provide a brief answer for each of the following questions in the space provided.

1. Explain why the process of test assembly, administration, and analysis is viewed as cyclical.

2. What is the suggested procedure for developing a test item file?

3. Compare and contrast what you would expect to be appropriate difficulty levels on a norm-referenced test that is expected to differentiate student performance, and a criterion-referenced mastery test that covers material all students should have learned.

4. Outline the general rules for formatting tests.

5. In conducting item analysis procedures, why is it suggested that you use the scores of 10 highest-performing and 10 lowest-performing students?

APPLICATION ACTIVITIES

Conducting a Basic Test Item Analysis

Chapter 11, among other things, introduced you to some of the basic techniques and purposes of conducting item analysis procedures on tests. This activity will help you to gain hands-on experience with item analyses by having you conduct some specific procedures with a simple example of an item from a multiple-choice test.

The following item was administered as part of a 35-item multiple-choice geography test given to three different classes of fifth grade students. The correct response (b) is indicated by an asterisk. As you can see, the response choices of the 10 highest and 10 lowest-scoring students are provided to the right of the test item.

| The capital of Nebraska is | High Scorers | Low Scorers |
|---|---|---|
| a. Kearney. | 1 | 2 |
| *b. Lincoln. | 8 | 4 |
| c. Omaha | 1 | 3 |
| d. California. | 0 | 1 |

Review the "Step-by-Step Item Analysis Procedure for Classroom Use" section in your textbook and analyze this item by computing the following:

1. Difficulty Level

2. Discrimination Index

3. Efficiency of Each Distractor Option

After you have computed these three things, try to come up with some suggestions for improving the test item, if you think any are warranted. You can check the basic answers to the three computations against the answers provided at the end of the chapter.

## ANSWERS TO GUIDED STUDY EXERCISES

### Definition Exercises

*Assembly*: The process of collecting or developing test items, reviewing them, formatting the test, and preparing test directions.

*Administration*: The process of providing instructions for test taking and then having the examinees complete the test within a given frame of time.

*Analysis*: The process of analyzing test results in order to make inferences about student performance and adequacy of the test.

*Test Item File*: A file of potential test items that are developed as you teach on a day-by-day basis.

*Practice Tests*: Tests given prior to the administration of a graded test. These tests are similarly formatted to cover content similar to the graded test, and allow the student to gain experience and familiarity prior to the actual test.

*Test Security*: The degree to which the actual contents of a test are not available to examinees prior to the test administration.

*Difficulty Level*: The percentage of students who correctly answer the test item.

*Coverage*: The breadth of material or concepts encompassed within a test.

*Item Analysis*: A variety of procedures used for analyzing individual test items as to their difficulty level, discriminating power, and technical adequacy.

*Indices of Discrimination*: Methods for indicating how well a given test items is able to sort or differentiate high-performing students from low-performing students.

## Multiple-Choice Questions

1. C
2. A
3. C
4. B
5. D
6. B
7. C
8. A

## Short-Answer Questions: Key Points

1. Test assembly, administration, and analysis is considered to be a cyclical process, as the three activities are all interrelated, and teachers do these activities in ongoing cycles.

2. The suggested procedure for developing a test item file is to develop item cards as lesson plans are developed, or at the end of each teaching session. The cards are then organized into a file and can be retrieved when it is time to develop the test.

3. With a norm-referenced test, it would be ideal to have about 50 percent of the class answer any test item correctly. With a criterion-referenced test, the ideal correct answer rate for individual items should be much higher--all students should get most items correct if effective instruction has taken place.

4. The general rules for formatting tests are:

   - Items with similar formats should be grouped together.
   - Proceed from the easiest to the most difficult items.
   - Make the test as readable as possible.
   - Avoid predictable response patterns.

5. Use of the 10 highest- and 10 lowest-scoring students is recommended in conducting item analysis procedures because the arithmetic procedures are simpler than in some other formulas, and this procedure produces results similar to those of more complicated formulas.

# APPLICATION ACTIVITY ANSWERS

## Conducting a Basic Test Item Analysis

1. Difficulty Level: 60%

2. Discrimination Index: .40

3. Efficiency of Each Distractor Option:

   Option a: Selected by one student in the high group, and two in the low group
   Option c: Selected by one student in the high group, and three in the low group
   Option d: Selected by no students in the high group, and one in the low group

# CHAPTER 12
# CONSTRUCTING AND USING DESCRIPTIVE MEASURES: QUESTIONNAIRES, INTERVIEWS, OBSERVATIONS, AND RATING SCALES

## CHAPTER SUMMARY

Most of the information and examples provided thus far in your text have been directly related to *cognitive* tests such as academic achievement and intellectual ability tests. You are introduced to a different angle on measurement and evaluation in Chapter 12--using measurement procedures to obtain various types of *descriptive* information. Specifically, you are introduced to four different types of descriptive measures that can be used for a wide variety of educational purposes: questionnaires, interviews, behavioral observation, and rating scales.

Chapter 12 begins with a brief overview on collecting descriptive data. Then you are led through major sections of the chapter dealing with each of the four methods of data gathering: questionnaires, interviews, behavioral observation, and rating scales. Each of these four major sections includes information that will provide you with the tools to develop appropriate objectives for that method, understand the appropriate uses and advantages of the method, understand the limitations of the method, and develop your own data-gathering system within that method.

By the time you have read Chapter 12 and completed the related activities in this workbook, you will have some new understanding of this specific area within educational measurement, and you will have been introduced to the basic "tools of the trade" in collecting descriptive information. Although this area of knowledge and skill can be highly useful for classroom teachers, it can also be very beneficial to you as a student. There is a good chance that you will be required to conduct a project in one of your education or social science courses that requires you to obtain descriptive information using one or more of the four methods described in the chapter. By using your text and this workbook as a guide, you will be able to complete these assignments in a more professional manner and make the learning experience more relevant to your goals as a potential educator.

## CHAPTER OUTLINE

Collecting Descriptive Data
Questionnaires
    Framing Questionnaire Objectives
    Constructing Questionnaire Items
    Questionnaire Format
    Pilot Testing the Questionnaire
    Analysis of Questionnaire Data

Interviews
    Specifying Interview Objectives
    Developing an Interview Schedule
    Guidelines for Constructing an Interview Schedule and Conducting the Interview
    Pilot Testing the Interview
    Analysis of Interview Data
Systematic Observation
    Observational Procedures
    Observational Objectives
    Observation Schedules
    Developing an Observation Schedule
    Methods of Recording Observational Data
    Pilot Testing
    Training the Observers
Rating Scales
    Some Problems with Rating Scales
    Rating Errors
    Rules for Developing Rating Scales

## GUIDED STUDY EXERCISES

### Definition Exercises

Locate the following terms in your text and briefly define each in the context of educational measurement and evaluation.

Closed-Form Items:


Open-Form Items:


Unstructured Interviews:


Fully Structured Interviews:


Semistructured Interviews:

Pilot Testing:

Duration Recording:

Frequency-Count Recording:

Interval Recording:

Continuous Recording:

Low-Inference Behavior:

High-Inference Behavior:

Observer Drift:

Error of Central Tendency:

Error of Leniency:

Halo Effect:

## Multiple-Choice Questions

Circle the letter of the item that best answers each question.

1. Closed-form items in questionnaires

    a. require more time and effort to complete than open-form items.
    b. are easier for the investigator to quantify and analyze than open-form items.
    c. should usually be a second choice to open-form items.
    d. tend to result in a higher refusal rate by respondents than open-form items.

2. Which of the following is the main problem in analyzing responses to open-form items on questionnaires and interviews?

    a. Assigning numerical values to the responses
    b. Obtaining appropriate information from the respondent
    c. Getting the respondent to complete the item in the first place
    d. Changing the response from quantitative to qualitative form

3. Compared with questionnaires, interviews tend to be

    a. less costly.
    b. more reliable.
    c. more valid.
    d. less objective.

4. Which of the following statements regarding semistructured interviews is *not* true?

    a. They are more flexible than fully structured interviews.
    b. Compared with fully-structured methods, they are a more objective and reliable method.
    c. They are the most useful method for most educational purposes.
    d. Most questions on them are open-format.

5. Which of the following is the best example of a high-inference behavior?

    a. Raising hand to ask teacher directions
    b. Getting out of seat
    c. Being mean to other students
    d. Working on task

6. Which of the following figures is recommended as the minimum acceptable amount of agreement between observers on a systematic observation schedule?

   a. 80%
   b. 85%
   c. 90%
   d. 95%

7. Due to John's behavioral problems, two of his teachers were asked to rate him using a standardized problem behavior rating scale. Because John excels in all academic subjects and is in a talented and gifted program, the ratings were influenced by his positive academic skills, and his behavior problems were not rated as severely as they should have been. What type of rating error occurred?

   a. Error of central tendency
   b. Error of leniency
   c. Halo effect
   d. Observer drift

8. Which of the following is an advantage of rating scales over observations and structured interviews?

   a. Higher reliability
   b. Higher agreement between raters
   c. More objective
   d. Easier to use

## Short-Answer Questions

Provide a brief answer for each of the following questions in the space provided.

1. Compare and contrast the uses and advantages of open-form and closed-form items.

2. What are some ways that you could minimize observer drift in a systematic observation system?

3. List one situation for each of the four types of data-gathering methods described in your text chapter (questionnaires, interviews, observation, rating scales) that would make it the method of choice.

4. What are some things that can be done to minimize the problems that often happen when using rating scales?

5. Compare and contrast unstructured, semistructured, and fully structured interviews. Give an example of when each of these would be the method of choice for gathering information.

## APPLICATION ACTIVITIES

### Deciding When to Use Specific Descriptive Methods

Now that you have been introduced to the four major methods of obtaining descriptive methods (questionnaires, interviews, observation, and rating scales), you have the basic background to understand how each of these methods is best utilized and what the limitations and problems associated with that method might be. This brief application activity will help you to further integrate your understanding of the four descriptive methods by having you make some decisions about which method would be most appropriate to solve some "real-life" educational problems.

The table on the following page lists six different educational scenarios where specific descriptive information is needed. Your task for this assignment is to carefully evaluate each scenario and then make a decision on which of the four descriptive methods would be best for that situation. Then you will need to provide some rationale or justification for the choice you made. Since your interpretation and rationale for each situation is bound to have some appropriate latitude, it is difficult to strictly specify "right" or "wrong" answers, but some examples of possible answers are provided at the end of this chapter for you to use as a guide.

| Educational Scenario | Method Selected | Rationale/Justification |
|---|---|---|
| A school board wants to obtain anonymous confidential information on the level and variety of illegal drug use by students | | |
| A school psychologist wants to obtain information on a referred student's behavior during instructional activities. | | |
| A teacher working on a master's thesis wants to obtain information on her students' attitudes toward school rules. | | |
| A teacher wants to obtain information on the study habits of his students prior to developing a study skills curriculum for them. | | |
| A researcher wants to obtain descriptive information on the type of social behaviors that occur in unstructured playground settings. | | |
| A principal wants to obtain information on her students' breakfast-eating habits prior to writing a budget for a school breakfast program. | | |

## ANSWERS TO GUIDED STUDY EXERCISES

### Definition Exercises

*Closed-Form Items*: Questionnaire items such as yes-no or multiple-choice options that do not allow for elaboration of responses and require less time and effort to complete.

*Open-Form Items*: Questionnaire items that allow the respondent to elaborate on the answer or to select an answer outside of any narrowly defined boundaries.

*Unstructured Interviews*: Interviews in which the interviewer is guided only by the broad objectives of the interview and does not need to follow a highly specific format.

*Fully Structured Interviews*: Interviews that maximize objectivity by allowing the interviewer very little latitude or leeway in asking or elaborating on interview questions; these interviews consist primarily of closed-form items.

*Semistructured Interviews*: Interviews in which the interviewer must follow a list of questions relating to the specific objectives that have been defined but has some latitude or leeway in gathering information; these interviews use primarily open-form questions.

*Pilot Testing*: The process of trying out an interview schedule, observation system, rating scale, or questionnaire on a small number of persons to work out problems and make improvements prior to implementing the system generally.

*Duration Recording*: A behavioral observation method where the observer watches for one or more specific behaviors and records the amount of time (or duration) that each behavior occurs.

*Frequency-Count Recording*: A behavioral observation method where the observer follows a list of specific behaviors to be observed on the observation schedule and enters a tally mark by it each time one of the targeted behaviors occurs.

*Interval Recording*: A behavioral observation method where the observer checks the target subject at regular specific intervals and records the behavior observed during that time period.

*Continuous Recording*: A behavioral observation method where the observer attempts to record all relevant behavior of the subject during the observation period.

*Low-Inference Behavior*: A behavior that requires very little judgment on the part of the observer to interpret.

*High-Inference Behavior*: A behavior that requires a great deal of judgment on the part of the observer to interpret.

*Observer Drift*: The tendency of different observers trained under the same observation system to independently and gradually change the definitions they are using to identify targeted behaviors.

*Error of Central Tendency*: The tendency of raters to choose responses near the center point on a rating scale when they are unsure of the correct rating.

*Error of Leniency*: The tendency of some raters to rate nearly everyone at the top of the scale, usually reflecting a reluctance to give anyone a "bad" rating on a rating scale item.

*Halo Effect*: The tendency of some raters to allow their general opinion of the subject being rated on the scale to influence their ratings on unrelated items.

## Multiple-Choice Questions

1. B
2. A
3. D
4. B
5. C
6. A
7. C
8. D

## Short-Answer Questions: Key Points

1. Closed-form items provide a highly structured format for completing questionnaires, require less time and effort on the part of the respondent, and are easier to analyze. Open-form items are easier to construct but are more difficult to respond to and to analyze.

2. Observer drift can be minimized by providing high-quality training prior to making the observations, through conducting periodic reliability checks with observers, and through adopting an observation system that is objective and easy to use.

3. 
   - Questionnaires: Best used for gathering specific concrete information at low cost
   - Interviews: Best used when complex and sensitive information must be obtained
   - Direct Observation: Best used for gathering objective setting-based information
   - Rating Scales: Best used for measuring general perceptions in a systematic manner

4. Some of the problems of rating scales can be minimized by obtaining composite ratings from several raters, using fewer levels of inference in the scale construction, and specifying a time period that the ratings are to be based on.

5. 
   - Unstructured interviews involve the interviewer being guided only by the broad objective of the interview--they are useful for assessing new or ill-defined problems.
   - Fully structured interviews allow the interviewer very little latitude, increase objectivity, and are useful for obtaining very specific information.
   - Semistructured interviews provide a structured format, but with some flexibility--they are the best choice for most educational information gathering.

# APPLICATION ACTIVITY ANSWERS

*Examples of possible responses--keep in mind that some scenarios could potentially be solved with more than one method.*

| Educational Scenario | Method Selected | Rationale/Justification |
|---|---|---|
| A school board wants to obtain anonymous confidential information on the level and variety of illegal drug use by students | Questionnaire | Confidentiality is essential, and the information to be obtained is specific |
| A school psychologist wants to obtain information on a referred student's behavior during instructional activities. | Observation | The desired information is specific to a particular setting and involves several variables |
| A teacher working on a master's thesis wants to obtain information on her students' attitudes toward school rules. | Rating Scale | Although interviews would also work, rating scales would provide similar standardized information and take less time to administer |
| A teacher wants to obtain information on the study habits of his students prior to developing a study skills curriculum for them. | Interview | Information that is both complex and sensitive can be obtained, and the teacher has maximum flexibility |
| A researcher wants to obtain descriptive information on the type of social behaviors that occur in unstructured playground settings. | Observation | Maximum objectivity is desired, and the information is specific to a setting |
| A principal wants to obtain information on her students' breakfast-eating habits prior to writing a budget for a school breakfast program. | Questionnaire | The information to be obtained is both specific and concrete, and since a large number of students are involved, the time and cost involved are important |

# CHAPTER 13
# GETTING IN TOUCH WITH STUDENTS' FEELINGS: MEASURING ATTITUDES AND INTERESTS

## CHAPTER SUMMARY

Have you ever taken a test or survey that asks you to respond to questions about your attitudes or interests? Chances are you have. The field of attitude and interest measurement is very large, and the types of measures involved range from ascertaining attitudes toward political candidates or issues to attitudes about shopping. The measurement of attitudes and interests within the field of education is just one small part of the entire range of attitude and interest measurement, and it is the focus of Chapter 13.

Chapter 13 begins with some discussion on the whys and hows of measuring affective outcomes and how attitudes influence behavior. A substantial portion of the chapter is then devoted to methods of constructing attitude measures, with a major emphasis on the use of Likert scales. Following this discussion, you are shown examples of three different types of attitude measures (school, curriculum, study) with six different specific measures overviewed. Chapter 13 ends with a section on measuring interests (with an emphasis on vocational interests), including general techniques for measuring interests and an overview of three different vocational interest measures.

Classroom teachers primarily focus their measurement and evaluation activities on academic performance. However, most teachers will have some occasions arise where the use of attitude and interest measurement will be beneficial, if not required. This chapter will help you to understand how attitude and interest measurement can be useful with your students and will even provide you with the tools you need to construct measures for you own purposes.

## CHAPTER OUTLINE

Measuring Affective Outcomes
The Influence of Attitudes on Behavior
Constructing Attitude Measures
    Measuring the Various Components of Attitude
    Methods for Developing Attitude Measures
    Steps in Developing a Likert Scale
    Advantages and Disadvantages of Likert Scales
    Reliability of Attitude Scales
    Validity of Attitude Scales
Typical Attitude Measures
    Measures of Attitude toward School, Teachers, and Specific School Subjects
    Measures of Attitudes toward Curriculum

    Measures of Study Attitudes
Measuring Interest
    Techniques for Measuring Interests
    Measures of Vocational Interest

# GUIDED STUDY EXERCISES

## Definition Exercises

Locate the following terms in your text and briefly define each in the context of educational measurement and evaluation.

Attitude:


Affective Measures:


Reactive Measures:


Likert Scale:


Item Analysis Techniques:


Interests:


Setting Influences:

**Multiple-Choice Questions**

Circle the letter of the item that best answers each question.

1. Compared with cognitive measures, affective measures tend to have

   a. lower levels of reliability and validity.
   b. higher levels of reliability and lower levels of validity.
   c. lower levels of reliability and higher levels of validity.
   d. higher levels of reliability and validity.

2. Terry, who is the athletic director at a large high school, believes that boys' and girls' athletic programs should receive equal amounts of funding and support. These beliefs are also shared by most of the school personnel and by most persons in the community. Regarding this issue, which of the following scenarios is most likely to occur?

   a. Terry's beliefs, feelings, and behavior will be inconsistent.
   b. Terry's beliefs and feelings will be consistent with each other, but not with Terry's behavior.
   c. Terry's beliefs, feelings, and behavior will be consistent.
   d. Terry's beliefs will be consistent with Terry's behavior, but Terry's beliefs and feelings will not be consistent.

3. Observing a few samples of an individual's behavior may not provide a very accurate estimate of his or her beliefs and feelings. What is the best explanation for this?

   a. People generally behave in ways inconsistent with their beliefs and feelings.
   b. People generally mask their true beliefs and feelings.
   c. Behavior tends to be consistent across settings, but not beliefs and feelings.
   d. Behavior can be greatly influenced by the setting.

4. Measuring attitudes generally emphasizes focusing on the cognitive and affective components rather than the behavioral components. Why is this true?

   a. Behavior cannot be effectively measured.
   b. Measuring behavior is more time consuming and expensive.
   c. Behavioral measurement is usually not allowed in school settings.
   d. Measuring behavior almost always provides useless information.

5. Jane has applied for a job as a night security guard in a store. She is required to take a test that asks a variety of questions about how honest she is, and whether or not she is ever tempted to take things that are not hers. Jane easily figures out what these questions are getting at and deliberately shapes her answers in order to make a favorable impression of an honest and trustworthy person. This is an example of

   a. a bipolar adjective checklist.
   b. a measure that is highly reactive.
   c. a reactionary test-taking style.
   d. a Likert scale.

6. Which of the following is the most widely used method of measuring attitudes?

   a. Bipolar adjective checklists
   b. Guttman scales
   c. Thurstone scales
   d. Likert scales

7. Which of the following is considered to be a disadvantage of Likert scales?

   a. They are difficult to adapt to measure different attitudes.
   b. Scoring is generally quite difficult.
   c. They depend on respondents' honesty or willingness to reveal true feelings.
   d. They can measure attitude intensity, but not attitude direction.

8. The Strong-Campbell Interest Inventory and the Kuder Occupational Interest Form both compare responses of subjects with those of

   a. persons in a variety of different occupations who are satisfied with their work.
   b. other students who are searching out careers they are potentially interested in.
   c. persons who previously worked in specific occupations but quit because they were unhappy.
   d. persons from similar ethnic and socioeconomic backgrounds.

## Short-Answer Questions

Provide a brief answer for each of the following questions in the space provided.

1. List and describe the three common components of attitudes.

2. What are some reasons for inconsistencies or discrepancies between an individual's attitudes and behavior?

3. How can attitudes simplify the day-to-day problems faced by people?

4. Why is a Likert or Likert-type scale generally considered to be a better way of measuring attitudes than the use of bipolar adjective checklists?

5. Compare and contrast the two interest inventories reviewed in the text--the Strong-Campbell Interest Inventory and the Kuder Interest Inventories.

## APPLICATION ACTIVITIES

### Analyzing a Likert Attitude Scale

One of the ways that a teacher can make a decision on whether or not to use specific tests is to evaluate the items in that test to see if they are of good technical quality. Refer to the section in your text ("Step 2: Write Statements for Your Scale") that lists Edward's general rules (a-n) on writing Likert scale items. Then, use Edward's criteria to evaluate the following items from the following brief hypothetical attitude scale. Each statement has at least one problem with it. Check your answers with the information provided at the end of this chapter.

## Attitude Toward My Teacher Scale

1. My teacher is fair to students; my teacher also knows
   a lot about the subject he/she teaches.                    SA  A  U  D  SD

2. My teacher drives a new car.                                SA  A  U  D  SD

3. My teacher had a good attitude toward students last year.   SA  A  U  D  SD

4. I always like my teacher.                                   SA  A  U  D  SD

5. My teacher is the worst teacher I have ever had.            SA  A  U  D  SD

6. Although my teacher is faced with many demands, and these
   demands are obviously quite difficult to simultaneously
   meet, I feel that my teacher does an exemplary job of
   balancing these competing demands and at the same time
   provides a positive and warm classroom environment.         SA  A  U  D  SD

# ANSWERS TO GUIDED STUDY EXERCISES

## Definition Exercises

*Attitude*: An enduring system of beliefs and feelings about an object, situation, or institution.

*Affective Measures*: Tests or other evaluation devices that measure constructs such as feelings, beliefs, and attitudes.

*Reactive Measures*: Attitude tests that contain items for which the respondent can determine the purpose and can structure his or her response to make a certain kind of impression.

*Likert Scale*: The most widely used scaling system on attitude scales--a method that allows the subject to respond to a statement by choosing one out of a range of several response options where a given response option is connected to a numerical score value.

*Item Analysis Techniques*: Methods of inspecting individual test items that are helpful in spotting and revising or removing items that are poorly written or do not have strong discriminating value.

*Interests*: Patterns of belief, feeling, and behavior that lead an individual to like, be indifferent to, or dislike certain activities.

*Setting Influences*: Situational influences on an individual's behavior that may produce circumstances where an individual may behave in ways that are not consistent with his or her attitudes.

## Multiple-Choice Questions

1. A
2. C
3. D
4. B
5. B
6. D
7. C
8. A


## Short-Answer Questions: Key Points

1. Attitudes are said to include cognitive (ideas and beliefs), affective (feelings), and behavioral (what one does) components.

2. One reason for inconsistencies between attitudes and behavior is that an individual's behavior toward an attitude object is closely related to the norms of behavior; that is, what other people think he or she should do or should not do. Thus, one may behave inconsistently with his or her attitude due to the pressure of social influence.

3. Attitudes can have the effect of simplifying day-to-day problems faced by people by creating a mind-set where one automatically thinks or acts a certain way rather than analyze each situation individually as it arises.

4. Likert or Likert-type scales are often used instead of bipolar adjective checklists because many paired adjectives are not really bipolar in nature, and midpoints between the adjectives are often not neutral.

5. Both the Strong-Campbell and Kuder interest inventories are widely used for helping persons make occupational choices. They both compare subject responses with responses of individuals working in specific occupations and divide occupations into categories based on different sets of occupational themes.


## APPLICATION ACTIVITY ANSWERS

Problems with "Attitude Toward My Teacher Scale" Based on Items from Edward's List:

1. i, l
2. d
3. a
4. j
5. e
6. g, h, i, l

# CHAPTER 14
# PICKING THE RIGHT YARDSTICK:
# ASSIGNING GRADES AND REPORTING STUDENT PERFORMANCE

## CHAPTER SUMMARY

Most of the chapters in your textbook deal directly with the process of testing. There is an underlying assumption that one of the primary purposes of using tests is to evaluate and grade student performance, but the topic of grading has thus far been covered only indirectly. Chapter 14 is a very specific treatment of this particular aspect of measurement and evaluation and provides you with background understanding and recommendations for the best practices in using grading systems.

Chapter 14 begins with a discussion of the history and background of grading systems in the United States prior to the 1900s, and then provides detailed information about different trends and movements in using grades from the early 1900s to the present time. A brief overview of the purposes and functions of grading is then provided, with a specific focus on how grades are used by students, parents, teachers and counselors, administrators, schools, and employers. Following this introduction and overview of grading systems and their uses, you are led on a comprehensive tour of types of grading systems, including types of comparisons and symbols used in grading, objections to using grades, and information on the predictive value of grades. Chapter 14 ends with a section that provides you with guidelines for assigning grades, which include specific recommendations on how to successfully use grades for educational systems, and the do's and don'ts associated with successful conferencing with parents of students.

The process of grading student performance does seem to take on slightly different functions and purposes at different grade levels, but even teachers of primary-age children will have the responsibility of evaluating their performance and then translating that evaluation into a permanent record of some kind. By carefully and thoughtfully considering the issues and information presented in Chapter 14, you can learn how to use the tools necessary for assigning grades and reporting student performance in a way that is compatible with your responsibilities and goals.

## CHAPTER OUTLINE

History and Background
Purposes and Functions of Grading
Types of Grading Systems
    Type of Comparison Used in Assigning Grades
    Type of Symbols Used
    Objections to Grades
    How Well Do Grades Predict Future Success?
Guidelines for Assigning Grades
    Developing and Reporting Grades

# GUIDED STUDY EXERCISES

## Definition Exercises

Locate the following terms in your text and briefly define each in the context of educational measurement and evaluation.

Curve-Based Grading:


Mastery Grading:


Achievement-Aptitude Grading:


Improvement-Based Grading:


Effort-Based Grading:


Contract Grading:


Letter Grading:


Percentage Grading:


Pass/Fail Grading:


Narrative Reports:

## Multiple-Choice Questions

Circle the letter of the item that best answers each question.

1. Which of the following statements about parent conferences is true?

   a. Limits should be placed on the amount of information presented to parents, as they may misinterpret it.
   b. Conferences are a good substitute for using grades.
   c. Student performance should be compared with that of other students during the conference.
   d. Parents should be treated as equals in the home-school relationship.

2. The basis for comparison in a grading system should

   a. usually be the same within a given school.
   b. differ within a school, but must be consistent across grades.
   c. be considered an individual classroom decision.
   d. be flexible from grade to grade within a school.

3. Which of the following practices for developing and reporting grades is considered to be a *poor* practice?

   a. Including opportunities for parent and student feedback within the grading system
   b. Mixing academic and nonacademic performance evaluation within a single letter grade
   c. Using the same basis for comparison within a given school
   d. Addressing *schoolwide* objectives within the grading system

4. The period of grade inflation during the 1960s and 1970s

   a. was accompanied by increases in student achievement.
   b. seems to have subsided.
   c. was a function of changes in the symbol systems used to assign grades.
   d. was surpassed by even greater grade inflation in the mid to late 1980s.

5. Which of the following statements regarding pass/fail grading is *false*?

   a. Pass/fail grading reduces student motivation.
   b. Pass/fail grading results in a decline in student achievement.
   c. Pass/fail grading causes lowered self-esteem in students.
   d. Pass/fail grading has not been found to be a good grading solution in most schools.

6. Which of the following statements about percentage grading is true?

   a. It has declined because of concerns that teachers could not effectively use it.
   b. It peaked in the 1920s and early 1930s, and then declined until about 1980.
   c. It is currently on the rise, due to problems with letter grade systems.
   d. It has been successfully challenged in court cases.

7. Grading on a curve is most reasonable when you are working with

   a. students of similar ability.
   b. small groups of students.
   c. large groups of students with a full range of ability.
   d. vocational or practical courses rather than academic courses.

8. For which of the following types of decisions are grades most useful?

   a. Short-term decisions
   b. Long-term decisions
   c. Administrative decisions that require detailed information
   d. Decisions relating to specific occupational placements

**Short-Answer Questions**

Provide a brief answer for each of the following questions in the space provided.

1. How have grading practices in schools changed since the early 1900s?

2. What is the evidence on how student competition for grades affects student performance?

3. What effect did the progressive education movement have on grading practices?

4. List the advantages and disadvantages of using pass/fail grading systems.

5. Of the seven objections to using grades listed in your text, which seems to be the most valid?

## APPLICATION ACTIVITIES

### Weighing the Pros and Cons of a Narrative Report System

The school board and administration of a large, urban school district have recently decided to turn one of the many high schools in the district into a "magnet" school where several different non-English languages would be taught, including Japanese, Russian, German, and Spanish. It is assumed that creating this special magnet site will revitalize the school and attract bright, motivated students from around the city. During the summer before the new magnet site is fully instituted, a group of teachers and administrators are assigned to serve as a task force for developing recommendations for school academic policies. These individuals are very excited about the new changes and want to help turn the school into a high quality "liberal arts" type of learning environment. The group discusses the proposal that rather than assigning absolute grades (either letter or pass/fail), teachers would grade student performance by collecting samples of student work and making brief narrative evaluative reports that will go on the student transcripts. The committee members are generally in favor of this idea and present it to the board for approval. Assume for a few minutes that you are a board member who has been given the task of evaluating this particular recommendation. The committee has enumerated the strengths and advantages of the new grading system. Now, you must take a more critical look at what problems might be involved. Make a list of the major concerns that you might find over this recommendation. Some examples of concerns are found at the end of this chapter.

## ANSWERS TO GUIDED STUDY EXERCISES

### Definition Exercises

*Curve-Based Grading*: The practice of grading student performance based on a comparison against performance of all students in the class.

*Mastery Grading*: The practice of grading student performance by determining the percentage of the material the student has demonstrated mastery of.

*Achievement-Aptitude Grading*: The practice of grading student performance based on how well the performance compares to the student's aptitude or ability.

*Improvement-Based Grading*: The practice of assigning grades to students based on how much improvement they demonstrate over previous performance.

*Effort-Based Grading*: The practice of grading student performance based on how hard the student tries.

*Contract Grading*: The practice of individually specifying student objectives and performance in a written contract and using completion of the contract as an absolute standard for grading student performance.

*Letter Grading*: The practice of grading student performance by assigning a letter grade (usually A-F), wherein each letter represents a designated level of performance.

*Percentage Grading*: The practice of evaluating student performance by assigning a number between 0 and 100 to represent the percentage of material a student has learned. This system was popular until the 1920s but is used less frequently today.

*Pass/Fail Grading*: The practice of grading students' performance based on whether or not they receive a minimum passing grade.

*Narrative Reports*: The practice of providing a written, qualitative description of a student's performance as a way of grading or evaluating the student.

## Multiple-Choice Questions

1. D
2. A
3. B
4. B
5. C
6. A
7. C
8. B

## Short-Answer Questions: Key Points

1. Some of the major changes in grading practices since the early 1900s are that grades are now based on more objective information than they once were, and teachers do not have the total authority over grading practices that they once did. Another change since the early 1900s has been the common use of a standard A-F letter grading system.

2. There is some evidence that some students perform better in situations where they compete for grades, but this evidence was from research with highly motivated graduate students. The evidence as to the effect of competition for grades in K-12 settings is unclear.

3. One of the main impacts on grading practices caused by the progressive education movement was the shift to letter grades. Another impact that may have impacted grading practices was an increased emphasis on the need for freedom and democracy in the classroom.

4. Pass/fail grading has been found to reduce student motivation and is correlated with a decline in student achievement.

5. No answer is provided for short answer question 5, as your particular answer will be opinion-based and cannot be compared against an absolute standard.

## APPLICATION ACTIVITY ANSWERS

### Weighing the Pros and Cons of a Narrative Report System

Your listing of concerns and potential problems with the committee recommendation for a new, narrative-based grading system might include the following:

- Even with an abbreviated narrative grading system, there is a good chance that the amount of paperwork generated would be enormous. Where and how would the materials be stored, and what difficulties would be encountered with sending them to other schools when students move?

- How would already busy high school teachers find the time to effectively produce narrative grades, given that this system is probably more time consuming than a traditional letter grade system?

- Since it is assumed that the new magnet school will attract bright and highly motivated students, many of whom will be college-bound, how will the narrative reports be used by college admissions officers who must make selection decisions?

- How could the information in the narrative reports be effectively translated so that both students and their parents would have a good understanding of the outcome of grading?

# CHAPTER 15
# AVOIDING BEING CAUGHT IN THE CROSSFIRE BETWEEN STANDARDIZED TEST SUPPORTERS AND ALTERNATIVE ASSESSMENT ENTHUSIASTS

## CHAPTER SUMMARY

The first few chapters of your text introduced you to some of the controversies surrounding the use of tests. Now, you get the chance to jump into the swirl of debate over test use. Text Chapter 15 and this section of the workbook not only point out many of the arguments and criticisms over test use but more importantly provide you with well-reasoned, rational information about these controversial issues, which will allow you to make informed choices about them.

The chapter begins with a list of questions commonly raised about the use of educational tests, provides some working definitions about different types of tests, and then lays out 13 of the most common criticisms of testing. Then, a countercritique is presented to show you that some of the complaints against using tests are just plain wrong, and some are not well conceptualized. However, do not assume that this chapter was written only to defend the use of tests. The next section of the chapter reviews 12 common misuses of tests and suggests ways of avoiding such problems. The chapter ends with a section on alternatives to traditional standardized testing. The potential for using various alternate testing systems is discussed, along with some of the problems and difficulties that using these alternative techniques involves.

By the time you have read Chapter 15 from the text and gone through the exercises and activities for it from this section of the workbook, you should be well acquainted with many of the problems and criticisms of testing, and you should also have a realistic idea of what can be done to minimize or overcome these difficulties. Gaining a good working knowledge of the potential pitfalls and abuses of testing will help you when you are responsible for selecting tests for different educational purposes. The section on alternative testing methods at the end of Chapter 15 provides you with the knowledge to make a realistic appraisal of how useful some of these techniques might be in your specialty area. Alternative assessment is currently a "hot" topic in education, and gaining a solid understanding of the uses, misuses, and problems associated with all kinds of educational testing will help you in making informed choices in the future.

## CHAPTER OUTLINE

Criticisms and Cautions Concerning Educational Tests
    Common Criticisms of Standardized Tests
Critiquing the Criticisms of Testing
    Standardized Tests Are Strongly Supported by Many Practitioners, Parents, and Policy Makers
    Most Criticisms of Testing Have No Empirical Basis
    Many Criticisms of Testing Assume Teachers and Other Test Users Are Naive and Credulous

Even Though Standardized Tests Are Imperfect, They Still Seem Somewhat Better Than the Alternatives
Most Criticisms Are Really Not Criticisms of the Tests But Rather of Their Misuse or Abuse

The Potential of Recently Proposed Alternatives to Standardized Testing
What Is Alternative Assessment?
Our Position Concerning Alternative Assessment
Critical Issues Facing Alternative Assessment
Criteria for Determining Your School's Readiness for Expanding the Use of Alternative Assessment
Potential Uses of Portfolio Assessment in Your School

## GUIDED STUDY EXERCISES

### Definition Exercises

Locate the following terms in your text and briefly define each in the context of educational measurement and evaluation.

Standardized Tests:


Norm-referenced Tests:


Criterion-referenced Tests:


Minimum Competency Tests:


Content Mismatch:


Test Bias:


Behavior Sample:

Test Misuse:


Alternative Assessment:



**Multiple-Choice Questions**

Circle the letter of the item that best answers each question.

1.  Which of the following is *not* one of the common criticisms of standardized tests?

    a. They do not promote student learning.
    b. They penalize bright, creative students.
    c. They are racially, culturally, and socially biased.
    d. They are used to hold teachers and school systems accountable.

2.  Which of the following potential problems of standardized tests reflects more of an indictment against the person(s) who selected it than the test itself?

    a. The test content is mismatched with the content of the curriculum.
    b. The content of the test is biased against a specific cultural group.
    c. The test violates the privacy of a student or family.
    d. Test anxiety interferes with a student's performance.

3.  What type of role do tests play in fostering student learning?

    a. An indirect role
    b. A direct role
    c. No role at all
    d. A negative role

4.  A careful analysis of standardized test items shows that large proportions of them measure

    a. only the ability to memorize information.
    b. relatively complex mental processes.
    c. trivial information.
    d. only left-hemisphere brain processes.

5. In a recent Gallup poll, 81 percent of parents surveyed described standardized tests as

   a. not useful.
   b. somewhat useful.
   c. moderately useful.
   d. very useful.

6. Which of the following statements is true concerning the criticisms of testing?

   a. They have generally been proven through research.
   b. They have been discredited by years of test use.
   c. There is a large amount of evidence to support them.
   d. There is little evidence to support them.

7. Effects of standardized testing on teachers have been found to

   a. confirm their judgments from observations and other information sources.
   b. shape their choice of curriculum materials.
   c. form their basic attitudes toward students.
   d. negatively affect their relationships with students.

8. Most of the criticisms of standardized tests

   a. are totally unreasonable.
   b. are really criticisms of their misuse or abuse.
   c. have received strong empirical support.
   d. are indictments against the test development industry.

## Short-Answer Questions

Provide a brief answer for each of the following questions in the space provided.

1. Compare and contrast norm-referenced and criterion-referenced tests.



2. Respond to the criticism that standardized tests usually measure only trivial information and require only low-level mental processes.

3. Briefly appraise the value of standardized tests versus student portfolio assessment.

4. What is the main problem with setting standards for minimum performance on tests?

5. What are some of the problems with incorporating alternative assessment methods in the schools?

## APPLICATION ACTIVITIES

Suppose that you are a classroom teacher who has been asked to be a member of a districtwide committee on testing and measurement issues. Your committee has the responsibility of making recommendations to the central administration and school board on any issue that involves testing or measurement across the school district. Your committee is currently faced with the need to make a decision regarding the use of an alternative assessment method in all high school English classes in the school district. What has been proposed is that the school district abandon the use of districtwide standardized achievement tests in English at the end of each year, and instead use evaluations of student work portfolios, which might include such things as written papers, projects, and assignments. What has spawned this proposal is the fact that your school district has a large percentage (approximately 30 percent) of students for whom English is not their primary language, or who are bilingual. There is a great deal of concern that for many students, the standardized tests currently given at year's end are not appropriate measures of their achievement, and may be culturally biased.

Your job is to investigate the merits of this proposal and bring back to the committee a list of pros and cons that you see associated with it. You will also need to make a recommendation to the committee as to whether this proposal should be adopted. Use the box on the next page as a worksheet. You can compare your ideas with some possibilities that are presented in the Application Activity Answers section.

| Issue: Whether or Not to Abandon the Use of Standardized English Achievement Tests at the High School Level in Favor of Evaluation of Student Work Portfolios ||
|---|---|
| **PROS** | **CONS** |
|  |  |

RECOMMENDATION TO THE COMMITTEE:_____

_____

_____

## ANSWERS TO GUIDED STUDY EXERCISES

### Definition Exercises

*Standardized Tests*: Tests that use a standard set of instructions for administration, scoring, and interpretation.

*Norm-referenced Tests*: Tests that compare a student's score with those of other students.

*Criterion-referenced Tests*: Tests that compare a student's test score with some absolute standard or criterion of success, but not with the performance of other students.

*Minimum Competency Tests*: Tests used to measure attainment of minimum standards of competency for key educational pass/fail decisions about individuals.

*Content Mismatch*: When the content of a test is not closely related to what is being taught.

*Test Bias*: A condition that exists when people of equal ability on what is measured by the test do not receive equal scores.

*Behavior Sample*: The limited range of behaviors elicited by a test, which usually represents only a portion of the entire domain of interest.

*Test Misuse*: Using tests for purposes other than those for which they are designed and validated.

*Alternative Assessment*: A wide array of assessment methods that have been developed as alternatives to standardized testing, that have been collectively referred to as authentic assessment, performance assessment, direct assessment, nontraditional assessment, and so forth.

**Multiple-Choice Questions**

1. D
2. A
3. A
4. B
5. D
6. D
7. A
8. B

**Short-Answer Questions: Key Points**

1. Norm-referenced tests are used to compare student performance with that of a particular reference group. Criterion-referenced tests are used to compare student performance against attainment of a specific set of criteria.

2. While it is true that standardized test questions can easily be written (and often are) to measure only trivial information and low-level mental processes, it is also true that with the correct planning, they can be developed to test many complex mental processes and to obtain very sophisticated information.

3. The main value of standardized tests is that they provide a regular or typical format for administration and for interpretation of results. In the case of standardized norm-referenced tests, they also provide a normative criterion on which to judge student performance. Student portfolios, on the other hand, would be a product that is unique for each student, both in terms of how they were designed and what information they provide. Student portfolios would probably be more time consuming and difficult to evaluate but would also provide a greater amount of qualitative information on student performance.

4. The main problem with setting standards for minimum performance on tests is the process of deciding what the standard should be. It is difficult to select a standard so that there can be a great deal of confidence in how effectively it acts as a "gate."

5. Some problems with incorporating various alternative assessment methods into schools include: (a) resistance among staff, (b) making the alternative assessment technique sufficiently clear, (c) training staff members in effectively using the alternative methods, (d) selecting methods that have clear outcomes for student performance, and (e) public resistance to the alternative methods.

## APPLICATION ACTIVITY ANSWERS

The following table provides some possible responses for the Application Activity exercise. The final recommendation as to whether or not to adopt the proposal is yours, so we have not made an ultimate suggestion here.

| Issue: Whether or Not to Abandon the Use of Standardized English Achievement Tests at the High School Level in Favor of Evaluation of Student Work Portfolios ||
|---|---|
| PROS | CONS |
| 1. The current standardized English achievement tests may not be good or fair measures for the large number of ESL or bilingual students. | 1. Evaluating student work portfolios will be very labor intensive. |
| 2. The use of student work portfolios will provide more information on individual student performance than do norm-referenced test scores. | 2. There is a potential problem of inconsistency between different evaluators of portfolios. |
| | 3. How will the results of the evaluations of student work portfolios be translated to parents and the school board? |
| 3. The use of student work portfolios is likely to have stronger instructional implications than norm-referenced test scores. | 4. The portfolios will tell us only about individual achievement, not group achievement. |
| 4. The requirements of the student work portfolio can be tailored to meet specific needs and objectives in this school district and with individual students. | 5. The school board and community will not be able to use this evaluation information to make comparisons between past and present performance, or between our school district and other districts. |
| | 6. The public and media have the expectation that we will administer standardized norm-referenced achievement tests and release average scores for the district. |

# CHAPTER 16
# FINDING AND SELECTING MEASURES THAT CAN HELP SOLVE YOUR EDUCATIONAL PROBLEMS

## CHAPTER SUMMARY

Have you ever been in the frustrating situation of trying to do a job without the right tools? Or, worse yet, trying to do a job and not even knowing what tools you need? Chapter 16 in your text is about tools--the kind of tools you need to become familiar with in order to find and select tests to help you solve educational problems. It is definitely a "hands-on" chapter that provides you with step-by-step directions for using several different resources to find and select appropriate educational tests.

Chapter 16 begins with a step-by-step outline of how to go about selecting educational measures. From defining and identifying the possible causes of the problem you are faced with to locating and searching for what measures are available to help you with the problem, this section provides you with the tools you need to locate appropriate measures. After you have located measures that you want to consider, the next step is to go about evaluating how well these measures will work for you, and then make a final selection. Thus, the second major section of Chapter 16 provides specific guidelines to follow in evaluation and selection of measures. These guidelines include evaluating the technical properties of the test and deciding how practical a test will be for your specific use.

There is a good chance that, if you have taken several previous courses in education or psychology, you have already had the experience of doing a test review using the *Mental Measurements Yearbook* or conducted a literature search using *Psychological Abstracts* or *Current Index to Journals in Education*. If you have, the text chapter and the related exercises in this workbook will help you to further understand and refine your skills for using these resources. If you have not, then the text chapter and related exercises in this workbook will provide you with a working understanding of how to go about basic library research in education and psychology.

## CHAPTER OUTLINE

Steps in Using Educational Measures
    Defining the Problem
    Identifying Possible Causes of the Problem
    Determining What Tests Are Available to Measure These Variables
    Locating Relevant Measures
    Conducting a Manual Search
    Conducting a Computer Search
Evaluating and Selecting Measures
    Test Validity

Test Reliability
Normative Data
Administration and Scoring
Clarity of Test Content
Cost of the Test

## GUIDED STUDY EXERCISES

### Definition Exercises

Locate the following terms in your text and briefly define each in the context of educational measurement and evaluation.

Variable:


*Mental Measurements Yearbook*:


ETS *Test Collection Bibliographies*:


*Current Index to Journals in Education (CIJE)*:


*Resources in Education (RIE)*:


*Psychological Abstracts*:


*Thesaurus of ERIC Descriptors*:


And-or connectors:

Proximity Search:



**Multiple-Choice Questions**

Circle the letter of the item that best answers each question.

1. Other things being equal, the larger the size of a norm sample, the more

    a. valid the test will be.
    b. accurately it represents the population from which it was selected.
    c. difficult it is to determine test reliability.
    d. variance is introduced into test scores.

2. Clarity of a test is best evaluated

    a. by looking at the information in the ETS *Test Collection Bibliographies*.
    b. through studying the test review in the *MMY*.
    c. by studying the test itself.
    d. through conducting an ERIC search.

3. The time required to determine the purpose of testing and type of test needed and to locate and review available tests is an example of which type of cost?

    a. Test developmental costs
    b. Test administration costs
    c. Test scoring costs
    d. Test dissemination costs

4. Which of the following is a shortcoming of the reviews of tests found in the *Mental Measurements Yearbook*?

    a. Critical evaluative reviews are not provided.
    b. Only brief descriptive information is provided.
    c. Only commercially published tests are reviewed.
    d. Most university libraries do not have the *MMY*.

5. Which of the following is a shortcoming of the reviews of tests found in the ETS *Test Collection Bibliographies*?

    a. Only brief descriptive information is provided.
    b. Only commercially published tests are reviewed.
    c. Very few university libraries have the bibliographies.
    d. They tend to provide overly favorable reviews of tests published by ETS.

6. Shana is attempting to find a list of possible tests to assess self-concept. She finds three relevant terms in the *Thesaurus of ERIC Descriptors* and then looks for the references that match these descriptors, one at a time, in the *RIE* and *CIJE*. What type of a search is Shana conducting?

    a. A limited computer search
    b. A computer search using *or* descriptors
    c. A manual search
    d. A random examination of professional journal contents

7. The advantage of using a computer search for locating specific types of journal articles is

    a. usually outweighed by the costs involved.
    b. that you can cross-reference descriptors and eliminate many irrelevant articles.
    c. that virtually every school in the nation is connected to a computer database.
    d. that most university libraries do not have sufficient resources for a manual search.

8. Victor begins a computerized literature search by using the terms *mathematics anxiety* and *secondary school students*, and joining them with the *and* connector. This method of term selection will result in

    a. only references with both descriptors being selected.
    b. references containing either descriptor being selected.
    c. only references that begin or end with both descriptors being selected.
    d. more references being found than if he used the *or* connector.

## Short-Answer Questions

Provide a brief answer for each of the following questions in the space provided.

1. Compare and contrast the type of information available in the *Mental Measurements Yearbook* and the ETS *Test Collection Bibliographies*.

2. In terms of number of references that are located in a computer search, what is a good rule of thumb for deciding if you should print the entire selection or to narrow it to a smaller number before printing the references?

3. When using a computerized literature search to locate specific information from educational journals, when is the use of a proximity search advised?

4. In general, what is considered to be the single best source for locating information on published tests, and why is this source said to be the best?

5. What characteristics should a teacher look for when examining the normative data in a test review or manual?

## APPLICATION ACTIVITIES

### Researching a Measure in Your Area

This application activity will presuppose that you have access to a college, university, or public library that has some of the basic research tools that were overviewed in the text chapter. Chances are that if you are currently taking a course in measurement and evaluation you will have access to at least some of these tools. If not, you will not be able to do this application activity.

This activity is essentially a series of steps for you to follow that will acquaint you with the basic methods of actually locating and evaluating information on an educational measure in your area. Since your own research will produce unique findings, no answers to this activity are located herein.

*Step One*: Pick an area within your educational specialty for which you would like to identify a test, or, in other words, identify a variable you want to measure. This can be either a cognitive or affective area. If you do not have a specialty area, simply identify an area that interests you.

*Step Two*: Refer to the latest edition of the *Thesaurus of ERIC Descriptors* in the reference section of your library, and locate at least one or two terms that are synonymous with or very close to the variable you want to measure.

*Step Three*: Look up the descriptor term(s) you selected from the *Thesaurus* in the most recent volume of *Current Index to Journals in Education*, and write down the "EJ numbers" assigned to each entry in the CIJE.

*Step Four:* Refer to each EJ number in the Main Entry Section of *CIJE*, and read the pertinent descriptors of the article, as well as the abstract.

*Step Five:* If the abstract of the article appears to describe a measure that is relevant to what you are looking for, and interests you, write down the pertinent information on the article, and locate the article in the periodicals section of your library. Repeat this step until you have located at least one or two measures that you think would help you measure the educational variables you are interested in.

*Step Six:* Using the Ninth or Tenth *Mental Measurements Yearbook*, locate a review of the measure(s) you have selected, and read through it.

*Step Seven*: Using the example of a test evaluation form from Chapter 16 of your text, rate the measure(s) you have identified. Use the criteria on the example form, as well as the information in the second major section of the chapter ("Evaluating and Selecting Measures"), and determine whether or not you think the measure(s) would be a good choice for you.

*Step Eight*: If possible, once you have identified a measure that you believe will meet your needs, try to actually locate and examine a copy of the measure. Most colleges or departments of education have resource libraries with some testing materials, and this would be a good place to start. You might also check with your course instructor to find possible sources for locating the actual test you have selected.

## ANSWERS TO GUIDED STUDY EXERCISES

### Definition Exercises

*Variable*: A characteristic or kind of performance that a test measures, something that "varies" from person to person.

*Mental Measurements Yearbook*: A series of "yearbooks" published at irregular intervals over the past 50 years that contain extensive critical reviews of commercially published tests. The most recent versions are the Ninth (published in 1985) and the Tenth (published in 1989).

*ETS Test Collection Bibliographies*: A series of over 200 bibliographies published by the Educational Testing Service. These bibliographies provide brief descriptions of over 11,000 commercially published and experimental tests.

*Current Index to Journals in Education (CIJE)*: A primary source for locating journal articles or other publications related to education. *CIJE* can be found in reference sections of most large libraries.

*Resources in Education (RIE)*: A primary source for locating journal articles and other publications in education, also found in reference sections of most large libraries. *RIE* is published by the Educational Resources Information Center (ERIC).

*Psychological Abstracts*: The principal primary source for locating journal articles and other publications in the field of psychology, also found in the reference section of most large libraries.

*Thesaurus of ERIC Descriptors*: A resource available in the reference section of most large libraries that provides a standard set of terms that can be used to classify and describe the contents of education publications. This resource is useful in locating keywords for a manual or computer search of the literature in a specific area.

*And-or connectors*: Methods of specifying how to connect keywords or terms in a computerized literature search. Using the *and* connector between specified terms tells the computer to locate only references that contain both descriptors joined by *and*. Using the *or* connector tells the computer to locate all the references that contain either one of the descriptors joined by *or*.

*Proximity Search*: A method used in a computerized search of the educational or psychological literature that scans all the words in the bank of memory to locate the presence of specifically identified terms.

## Multiple-Choice Questions

1. B
2. C
3. A
4. C
5. A
6. C
7. B
8. A

## Short-Answer Questions: Key Points

1. The *Mental Measurements Yearbooks* (*MMY*) are a series of comprehensive test reviews that have been published over the past five decades. The most recent edition is the Tenth. The MMY is found in most university libraries. The ETS *Test Collection Bibliographies* are more exhaustive than the *MMY* in terms of number of tests evaluated, but provide much less information about the tests.

2. It is recommended that if a computer search produces more than 50 references, you either instruct the computer to select the most recent 50 or else narrow down the focus of the search to produce fewer, but more specific, references.

3. A proximity search is advised when you are researching recently discovered or unusual variables that may not be listed in the key terms descriptors.

4. The *Mental Measurements Yearbooks* are said to be the single best source for locating information about published tests because (a) a large number of tests are included in the reviews and (b) the evaluative information is much more comprehensive than what is normally found in the ETS bibliographies or in journal articles.

5. Teachers evaluating the normative data of a test should consider how well the norm group represents their particular students, as well as how large and representative the test norms are for the general U.S. population.

# CHAPTER 17
# WHAT HAVE MY STUDENTS LEARNED?
# AN INTRODUCTION TO STANDARDIZED ACHIEVEMENT MEASURES

## CHAPTER SUMMARY

The first several chapters of your textbook have referred frequently to various types of academic achievement tests. Chapter 17 will help you to explore the differences, similarities, and uses of academic achievement tests in much greater depth. The text chapter states that "bettering students' academic achievement is generally considered the primary goal of the public schools." With that statement in mind, it becomes quite apparent that gaining a good working knowledge of types and uses of academic achievement tests will help you to reach the most important goals of teaching, as well as to become an informed consumer of testing materials.

The chapter begins with an overview and description of six different types of achievement measures, including teacher-made, curriculum-embedded, diagnostic, single-subject, standardized, and tests for use with special student groups. Next, the chapter guides you through the ways that achievement tests are best used, from the perspective of teachers, administrators, and counselors. Along with this exploration of different uses, you are cautioned about some of the problems in interpreting standardized achievement test data, and you are introduced to the steps involved in actually developing standardized measures. Chapter 17 winds up with a detailed look at how a typical standardized academic achievement battery is utilized, using the Comprehensive Test of Basic Skills as an example. The chapter details how this test is used in actual practice and provides examples of the types of information yielded from the CTBS.

As was stated in the introduction to section three of the text, learning how not to "reinvent the wheel" will be an important skill in becoming an informed consumer of testing materials. Developing a basic working knowledge of the types and uses of academic achievement measures will enable you to perform better as an education professional. This knowledge is highly useful, if not critical, for teachers, administrators, and support personnel alike.

## CHAPTER OUTLINE

Types of Achievement Measures
    Teacher-made Achievement Measures
    Curriculum-embedded Achievement Measures
    Diagnostic Tests
    Single-Subject Achievement Measures
    Achievement Measures for Special Student Groups
    Standardized Achievement Test Batteries
Using Educational Achievement Data
    Teacher Uses of Achievement Data

Administrative Uses of Achievement Data
Counseling Uses of Achievement Data
Cautions in Interpreting Standardized Achievement Test Data
Steps in Developing a Standardized Achievement Test
A Look at a Widely Used Standardized Achievement Test
The Comprehensive Tests of Basic Skills (CTBS)

## GUIDED STUDY EXERCISES

### Definition Exercises

Locate the following terms in your text and briefly define each in the context of educational measurement and evaluation.

Teacher-made Tests:


Curriculum-embedded Measures:


Diagnostic Tests:


Single-Subject Tests:


Standardized Achievement Test Batteries:


Special Achievement Measures:


Content Analysis:


Curriculum Emphasis:

Chance Scores:


Base Scores:


Functional Level Testing:




**Multiple-Choice Questions**

Circle the letter of the item that best answers each question.

1. What is considered to be the primary goal of the public schools?

    a. Developing good citizens
    b. Bettering students' academic achievement
    c. Increasing students' self-esteem
    d. Providing specific skills training for jobs

2. Which of the following statements is true regarding testing at the functional level?

    a. More children are likely to score below chance or above ceiling levels.
    b. High-achieving students are likely to become bored.
    c. All students experience some success.
    d. Low-achieving students are more likely to be frustrated.

3. The Spanish/English Reading and Vocabulary Screening Test (SERVS) is designed to

    a. determine a student's dominant language.
    b. ascertain the degree of a student's acculturation in Anglo society.
    c. be used primarily by Spanish-dominant students.
    d. replace the CTBS Espanol as a basic Spanish language academic test.

4. Which of the following factors would be most important to consider in selecting a standardized achievement measure?

   a. Content validity
   b. Availability of alternate test forms
   c. Inclusion of subtest scores
   d. Criterion-related validity

5. Which of the following is *not* an example of a typical teacher use of educational achievement data?

   a. Organizing student learning groups
   b. Diagnosing student learning difficulties
   c. Planning instruction
   d. Evaluating program and curriculum

6. Concerning test reliability, what would normally be true of a comparison between an entire achievement test and a specific content objective?

   a. The specific content objective would have the same reliability as the total test.
   b. The reliability of the total test would tend to be lower.
   c. The reliability of either would be based on the content validity of the test.
   d. The reliability of the specific content objective would tend to be lower.

7. In developing annual goals and short-term instructional objectives as part of a disabled student's Individualized Educational Plan, standardized achievement measures should

   a. be the sole criterion used.
   b. not be used.
   c. be combined with other sources of information.
   d. be used only if grade-equivalent scores are utilized.

8. The Educational Testing Service reports that developing an entirely new test requires about

   a. one year of time.
   b. two years of time.
   c. three years of time.
   d. four years of time.

## Short-Answer Questions

Provide a brief answer for each of the following questions in the space provided.

1. Compare and contrast the type of information provided by curriculum-embedded achievement measures and diagnostic tests.

2. What are some ways that a teacher can compare the content of a standardized achievement test with the content of the instructional curriculum?

3. What are some appropriate and inappropriate uses of standardized achievement tests in developing Individualized Educational Programs for students with disabilities?

4. Provide three possible explanations for a school district obtaining very low districtwide scores in a single subject area.

5. Develop arguments both for and against standardized achievement test scores as a basis for evaluating teacher performance. Which argument seems to have the most merit?

## APPLICATION ACTIVITIES

### Evaluating Content Validity in Your Specialty Area

This suggested application activity is another "hands-on" activity that will require you to actually get out and evaluate educational materials, but will not provide you with a list of correct answers.

This activity will provide you with experience in evaluating standardized achievement tests in your specialty area that may prove to be very valuable when you are faced with the task of having to select and evaluate tests for actual use with your students.

In order for you to do this activity, you will need to locate two things. The first resource you will need to have will be curriculum materials or an instructional unit from your educational specialty area (if you have not yet selected a specialty area, choose one that you have some interest in for the purpose of this activity). Next, locate either a single subject achievement test for the area you have selected, or locate a standardized achievement test battery that includes coverage of the area you have selected. These resources might be most easily located at your college or university library, in a resource center of a department or college of education, or in the materials center of your local public school district. If you suspect that you will have difficulty locating these two materials, check with your course instructor for suggestions.

Once you have located the curriculum materials and the achievement test to be considered, go through the following steps to evaluate how well the test corresponds with the content of the curriculum:

1. First, simply become familiar with the curriculum materials, and then inspect each item from the test to see if the concept or information covered is also adequately covered in the curriculum. What percentage of items in the test actually correspond to content coverage in the curriculum?

2. Next, use the content analysis form provided in Chapter 11, and go through the suggested steps of a simplified content analysis that are listed in the text. Use the examples provided in the text as a guide.

3. After you have completed these tasks, respond to the following questions:

   a. Is the use of the test appropriate to measure what is taught in the curriculum?

   b. If the answer to question 3a is "no," what needs to be done to make the test more suited to the curriculum?
   c. What suggestions do you have for alternative ways to test the content of the curriculum you selected?

## ANSWERS TO GUIDED STUDY EXERCISES

### Definition Exercises

*Teacher-made Tests*: The most widely used type of tests, those that are developed by the teacher.

*Curriculum-embedded Measures*: Testing materials that are included or "embedded" in commercially produced instructional materials, which are often criterion-referenced and used to determine whether or not students have mastered specific curriculum objectives.

*Diagnostic Tests*: Measures that are typically criterion-referenced, focused on lower-achieving students, and designed to provide extensive information on individual student performance.

*Single-Subject Tests*: Academic achievement tests designed for assessing student performance in one area (e.g., math, reading, history).

*Standardized Achievement Test Batteries*: Academic achievement measures that utilize standard instructions, time limits, and materials and are designed to expose each student to the same testing situation in order to provide comparable information and minimize measurement errors.

*Special Achievement Measures*: Academic achievement measures designed for specific populations and purposes, such as for use with students with disabilities, bilingual students, or talented and gifted students.

*Content Analysis*: A systematic inspection of the content of a test, usually done to determine how well the content of a test matches the curriculum for the subject that is being tested.

*Curriculum Emphasis*: The extent to which a given academic curriculum emphasizes specific sets of facts, concepts, and skills.

*Chance Scores*: The probability on multiple-choice test items that a student can get a certain percentage of responses right simply by guessing. If there were four alternatives on a multiple-choice test, a student answering randomly would, on the average, have a chance of getting 25 percent of the items correct.

*Base Scores*: The lowest score level that can be obtained on standardized achievement tests.

*Functional Level Testing*: A newer method of administering standardized achievement tests such as the CTBS that uses "locator tests" to place students at a difficulty level where they can experience some success and less frustration during the testing process.

## Multiple-Choice Questions

1. B     5. D
2. C     6. D
3. A     7. C
4. A     8. B

## Short-Answer Questions: Key Points

1. Curriculum-embedded achievement measures are brief tests that are included for each unit of a curriculum so that a teacher can frequently monitor student progress. Diagnostic tests, like curriculum-embedded measures, are usually criterion-referenced. They focus on the low-achieving student rather than measuring the whole educational range, and measure learning objectives in greater detail and depth than do curriculum-embedded measures.

2. The only way to establish the content validity of a standardized test for specific uses is to conduct a content analysis of the objectives covered in both the local curriculum and the standardized test. Conducting a content analysis in this manner involves a number of steps, including (a) analyzing the curriculum for each subject and grade level, and listing the broad content objectives, (b) reading each test item and listing them alongside the curriculum objectives, and (c) deciding which objectives are adequately covered, and which are not.

3. The appropriate uses of standardized achievement tests in developing Individualized Educational Programs include determining the student's present level of educational functioning and determining whether short-term instructional objectives are being met. It is inappropriate to use standardized achievement tests as the sole basis of information for developing and refining Individualized Educational Plans.

4. If students from an entire school district obtain very low average scores in a single subject area, then three explanations are possible: (a) the curriculum in that district may be poorly matched to the test, (b) the test may be poorly constructed, or (c) students in that district may be low in ability in the measured area.

5. An example of an argument favoring the use of standardized achievement test scores as a basis for evaluating teacher performance is that increasing student academic achievement is the single most important aspect of the teacher's role. An argument against this use of standardized achievement test scores is that there are factors (such as socioeconomic status and level of parent involvement) that contribute to student achievement but are beyond the control of the teacher.

# CHAPTER 18
## KEEPING YOUR FINGER ON THE STUDENT'S PULSE: USING TESTS TO DIAGNOSE STRENGTHS AND WEAKNESSES

## CHAPTER SUMMARY

You should be aware at this point in your study of educational measurement and evaluation that academic tests are not all designed for similar purposes. Most of the information you have been exposed to at this point has dealt with academic achievement tests, which are most often designed to provide a general measure of a student's level of attainment in a given academic subject. Although academic achievement tests can do a good job of telling you what level of attainment your students have reached, they most often do not provide you with specific information on what specific academic skills they have mastered or not mastered, pointing to the need for additional review. This is where we use diagnostic tests, which is the subject of Chapter 18 in your text.

The chapter begins with a description and overview of what diagnostic tests are, how they are used, and how they are best selected. Then a large section of the chapter is devoted to providing detailed descriptions of some types of diagnostic tests that are typically used. The chapter focuses on examples of diagnostic tests used for reading (Stanford Diagnostic Reading Test) and math (Diagnostic Math Inventory and Mathematics System), since these are the two areas that are most commonly measured in diagnostic testing. Detailed descriptions and reports are provided for these two examples, with suggested learning activities that will help you better understand how they are used. The end of the chapter is devoted to an exploration of how diagnostic tests are used with exceptional students, and you are introduced to the use of computerized adaptive testing, adaptive behavior tests, and specialized diagnostic tests that are designed to be used with students who have disabilities.

Unlike many of the previous chapters, Chapter 18 does not introduce you to a large number of new technical terms and concepts. This chapter allows you to draw upon the technical knowledge you have already acquired on testing to help you understand how tests can be used to diagnose and help remediate specific learning problems of students. If you plan on working with primary-age students, low-performing students, or students with disabilities, you will find this topic to be especially relevant and useful. Even if you do not plan to work extensively with students in these groups, learning to understand how diagnostic tests work can be very useful to you, since most teachers will ultimately be members of multidisciplinary child study teams to help serve the needs of students with special learning problems in their school.

## CHAPTER OUTLINE

What Are Diagnostic Tests?
Selecting Diagnostic Tests

Diagnostic Reading and Mathematics Tests
    Stanford Diagnostic Reading Test (SDRT)
    The Diagnostic Math Inventory Mathematics Systems (DMI/MS)
    The Utility of Diagnostic Math and Reading Systems
    Computerized Adaptive Testing
Standardized Achievement versus Diagnostic Tests
Diagnostic and Screening Measures for Students with Disabilities
    Tests of Adaptive Behavior
    Assessment of Academic Performance of Students with Disabilities
    Scholastic Aptitude Measures for Students with Disabilities

## GUIDED STUDY EXERCISES

### Definition Exercises

Locate the following terms in your text and briefly define each in the context of educational measurement and evaluation.

Diagnostic Tests:

Linking Diagnosis to Remediation:

Computerized Adaptive Testing:

Adaptive Behavior:

Screening:

Individually Administered Tests:

SOMPA:

Phonetic Analysis:

Objectives Mastery Report:

**Multiple-Choice Questions**

Circle the letter of the item that best answers each question.

1. Which of the following statements concerning diagnostic tests is *not* true?

   a. They are able to identify the underlying cause of a student's difficulty.
   b. They are able to identify the specific nature of a student's difficulty.
   c. They are aimed primarily at low-achieving students.
   d. They focus on single subjects and narrow achievement ranges.

2. Which of the following is an advantage of using individually administered diagnostic tests?

   a. They require more training to administer.
   b. They are complex to interpret.
   c. They permit behavioral observations of test performance and emotional behavior.
   d. They can be given in less time than group-administered tests.

3. Individually administered diagnostic tests are usually administered by

   a. reading specialists or school psychologists.
   b. regular classroom teachers.
   c. school counselors.
   d. classroom volunteers.

4. The Stanford Diagnostic Reading Test was developed based upon the view that reading is a developmental process that can be broken into four major components. Which of the following is not one of these components?

   a. decoding
   b. vocabulary
   c. reading rate
   d. visual scanning

5. Phonetic Analysis, one of the subtests of the Stanford Diagnostic Reading Test, is said to measure

   a. the student's oral fluency rate.
   b. the student's ability to relate sounds and letters.
   c. the student's ability to decipher lexical error patterns.
   d. the student's ability to comprehend the meaning of what was read.

6. The Diagnostic Math Inventory has evolved into a system designed to provide an integrated program for mathematics assessment and instruction. *Based on this capability alone*, the test is best classified as a(n)

   a. norm-referenced test.
   b. domain-referenced test.
   c. objectives-referenced test.
   d. criterion-related test.

7. Which of the following is a drawback to using diagnostic systems like the DMI or SDRT?

   a. They require extensive study for a teacher to make effective use of them.
   b. They can be administered only by school psychologists.
   c. They have good face validity, but their technical properties are questionable.
   d. They require expensive computer hardware and software in order to be used.

8. Diagnostic tests such as the Social and Prevocational Information Battery and Skills for Independent Living Resource Kit are most often used with

   a. students transitioning out of high school.
   b. students preparing to enter elementary school.
   c. gifted and talented students.
   d. students with disabilities.

## Short-Answer Questions

Provide a brief answer for each of the following questions in the space provided.

1. What are some ways that diagnostic tests and achievement tests differ?

2. What is phonetic analysis, as measured by the Stanford Diagnostic Reading Test?

3. Why is computerized adaptive testing capable of decreasing testing time when compared with traditional methods of diagnostic testing?

4. List some categories of daily living skills that are measured by tests of adaptive behavior such as the SPIB.

5. Your text states that most diagnostic tests are administered individually, whereas most achievement tests are group-administered. What are the advantages and disadvantages of individual versus group test administration?

## APPLICATION ACTIVITIES

### Understanding How Diagnostic Tests Work

Essentially, diagnostic tests are very much like achievement tests, except that the content of diagnostic tests is developed so that it can be extensively analyzed and recorded to provide teachers with specific information on areas of strength and weakness in their students. In looking at extensive commercially produced diagnostic tests such as the SDRT or the DMI/MS for the first time, it may be difficult to see how the tests were put together in a way that really makes them "diagnostic." Therefore, this application activity will give you an opportunity to analyze the content of a simplified example of a diagnostic test in order for you to determine exactly what skill areas it measures.

The following 18 items are examples of basic mathematics problems that could be used in a diagnostic test for primary-grade students. Your task for this activity is to analyze the content of each of the problems and categorize each one as to the specific math skill(s) for which mastery is required. Make a list of skill categories, with the item numbers that belong in each category. A hint--there are six different categories, with each category having three items in it. Compare your work with the answers provided at the end of this chapter.

1. 2 + 6
2. 4 + 3
3. 7 + 5
4. 4 - 2
5. 7 - 1
6. 9 - 4
7. 11 + 13
8. 22 + 41
9. 32 + 55
10. 19 + 28
11. 64 + 57
12. 29 + 93
13. 14 - 11
14. 79 - 55
15. 87 - 13
16. 22 - 17
17. 71 - 29
18. 55 - 38

## ANSWERS TO GUIDED STUDY EXERCISES

### Definition Exercises

*Diagnostic Tests*: Tests designed to provide a precise measure of students' performances in a given subject area.

*Linking Diagnosis to Remediation*: The process of taking information obtained through diagnostic testing and using it to develop useful educational strategies.

*Computerized Adaptive Testing*: A method of selecting and administering diagnostic academic test questions to students that is done through the use of a computer. CAT has the advantage of cutting down testing time by 50 percent to 70 percent and at the same time providing more precise diagnostic information by individually basing test question selection on the student's level of competence on the testing tasks.

*Adaptive Behavior*: The skills needed for everyday independent living. Adaptive behavior tests are often administered to students with disabilities, particularly those with low cognitive skills.

*Screening*: The process of using tests in "narrowing the field" of students who should be given further assessment for specific purposes. Screening tests are typically used to identify students who are "at risk" of not being able to succeed in school, and the students identified in the screening process are often then administered a more comprehensive battery of diagnostic tests.

*Individually Administered Tests*: Diagnostic tests that are administered to one student at a time and require continual interaction with the examiner. These tests tend to be more complex to administer than group administered tests and are usually administered by trained assessment specialists.

*SOMPA*: The System of Multicultural Pluralistic Assessment, a comprehensive diagnostic test battery used for assessing exceptional students. It measures adaptive behavior, cognitive abilities, and sensorimotor performance.

*Phonetic Analysis*: A reading process analyzed on many diagnostic reading tests whereby correspondence between sounds and letter symbols is used to decode words.

*Objectives Mastery Report*: A report form utilized with the Diagnostic Math Inventory and Mathematics System that provides a list of the specific instructional objectives that a student has mastered or not mastered, and a list of those for which additional review by the student is needed.

## Multiple-Choice Questions

1. A
2. C
3. A
4. D
5. B
6. C
7. A
8. D

## Short-Answer Questions: Key Points

1. Diagnostic tests are used for identifying specific areas of academic performance deficit. Achievement tests are typically designed to provide an indication of how well the student has mastered the material in general, but they do not provide the level of specific detail for which diagnostic tests are designed.

2. Phonetic Analysis, one of the tests of Stanford Diagnostic Reading Test, measures a student's ability to relate sounds and letters. On this test, consonant and vowel sounds are related to their most common spellings, and students are also asked to give the ending sounds of various words.

3. Computerized adaptive testing is capable of reducing test administration time by 50 percent to 70 percent over paper and pencil tests because more precise measures are provided, which reduce the number of items that are administered.

4. Categories of adaptive behavior from the Social and Prevocational Information Battery (SPIB) include economic self-sufficiency, employability, family living, personal, and communications.

5. Individually administered tests are more costly to administer than group tests, in terms of time. However, they usually provide a level of detailed information and observational data that is not possible with group-administered tests.

## APPLICATION ACTIVITY ANSWERS

| Division of Basic Mathematics Skill Areas by Item | |
|---|---|
| SKILL AREA | ITEMS |
| one-digit addition | 1-3 |
| one-digit subtraction | 4-6 |
| two-digit addition | 7-9 |
| two-digit addition with carrying | 10-12 |
| two-digit subtraction | 13-15 |
| two-digit subtraction with borrowing | 16-18 |

# CHAPTER 19
# ASSESSING YOUR STUDENTS' POTENTIAL:
# A LOOK AT APTITUDE AND READINESS MEASURES

## CHAPTER SUMMARY

In the previous chapters of your text, there have been many references to such things as *aptitude*, *ability*, and *intelligence*, but no in-depth discussion of these terms. Chapter 19 provides you with information about the purpose and structure of aptitude and intelligence tests. You have probably taken several aptitude or intelligence tests by this point in your life, and you may not even be aware of how these tests differ from the many achievement tests you have taken. For example, you probably had to take an academic aptitude test to get into college (such as the ACT or SAT), and you may have taken a group intelligence test while you were in elementary or secondary school. This chapter will help you come to a better understanding of what these tests really measure, and how to best use them.

The chapter begins with a section discussing aptitude tests, including definitions of aptitude, uses of aptitude tests, and information on how aptitude tests are used with culturally different children. The chapter then provides a comparative discussion on the differences and similarities between achievement tests and aptitude tests. Next, you are provided with information on how to tell whether an aptitude test is a good measure. The chapter ends with a discussion of six different kinds of aptitude tests, ranging from intelligence tests to specific aptitude tests.

Aptitude testing has been a controversial topic in education for many years, and there will likely be continued heated discussion on this topic. You may have already had the opportunity to engage in discussions about the use of aptitude or ability tests in other education classes, and you may be in a position in the future where you will have some influence on how aptitude tests are used with your students. One of the main points to keep in mind as you complete this chapter and develop a basic working knowledge of this topic is that it is the *misuse* of aptitude tests that has created problems in the past. For the most part, the tests themselves have not been the culprits in any educational malpractice. Become familiar with the uses and issues surrounding aptitude tests, and you will be in a better position to make informed judgments in the future.

## CHAPTER OUTLINE

The Use of Aptitude Tests
    What Is Aptitude?
    What Are Aptitude Tests Used For?
    Measuring Aptitudes for Children from Different Cultures
Comparing Achievement and Aptitude Tests
    Achievement Tests
    Aptitude Tests

Summary of the Differences Between Achievement and Aptitude Tests
Characteristics of Good Aptitude Tests
Examples of Different Types of Aptitude Tests
    Intelligence Tests
    Multiple Aptitude Batteries
    School Readiness Tests
    Tests of Creativity
    Tests of Musical and Artistic Aptitudes
    Professional and Vocational Aptitude Tests
    Specific Aptitude Tests

## GUIDED STUDY EXERCISES

### Definition Exercises

Locate the following terms in your text and briefly define each in the context of educational measurement and evaluation.

Aptitude Tests:


Culture-Free Tests:


Predictive Validity:


Intelligence Tests:


Multiple Aptitude Tests:


School Readiness:


Creativity:

**Multiple-Choice Questions**

Circle the letter of the item that best answers each question.

1. For school-aged children, intelligence scores account for approximately how much of the variance in achievement test scores?

    a. 10%
    b. 25%
    c. 60%
    d. 75%

2. Measures of intelligence become relatively stable by

    a. age 1.
    b. age 6.
    c. age 12.
    d. age 18.

3. Because even the best measures of aptitude are imperfect, they

    a. should be used in conjunction with other information.
    b. should not be used to make decisions about individuals.
    c. should be used only in conjunction with individually administered IQ tests.
    d. should not be used to help determine admission into schools or vocations.

4. In comparison with aptitude tests, achievement tests

    a. have lower reliability.
    b. are shorter to administer.
    c. measure causes of problems more effectively.
    d. measure more specific types of outcomes.

5. The evidence to date on the Culture Fair Intelligence Test suggests that it

    a. is no better a measure than other widely used intelligence tests.
    b. effectively eliminates cultural influences on test scores.
    c. results in significantly higher scores for minority students.
    d. reduces the standard error of measurement on intelligence tests for minority students.

6. Which of the following is *not* an advantage of using aptitude tests over achievement tests?

   a. They are quick and economical to administer.
   b. They can be used before any instruction occurs.
   c. They are appropriate for students from a variety of backgrounds.
   d. Aptitude tests have better content validity than achievement tests.

7. Individually administered intelligence tests are most often used for which of the following purposes in educational settings?

   a. To help decide whether children need special education services
   b. To develop curriculum-linked instructional recommendations
   c. To place students in talented and gifted programs
   d. To evaluate the effectiveness of instructional programs

8. Intelligence tests for infants tend to

   a. always have poor reliability.
   b. have little predictive validity.
   c. produce scores that are as stable as adult intelligence test scores.
   d. successfully predict future performance.

## Short-Answer Questions

Provide a brief answer for each of the following questions in the space provided.

1. Describe the relationship between aptitude and achievement.

2. How successful have efforts been to remove the influence of cultural variables from aptitude tests?

3. Compare and contrast the terms *aptitude* and *intelligence*.

4. How well does the Graduate Record Exam (GRE) predict performance in graduate school?

5. Why is measuring creativity so problematic?

## APPLICATION ACTIVITIES

### Deciding What a Test Really Measures

Suppose that a group of measurement and evaluation specialists has spent several years developing and field testing a new intelligence test, the Culture-Free Intelligence Scale (CFIS). The goal in developing this test was to produce a general cognitive ability test that was free of cultural influences and thus a fairer test for use with minority group members. The initial research with the CFIS has suggested that it has levels of test reliability that are equal to the best intelligence tests in use. Now, it is time to go through the process of determining how valid the test is. The researchers have produced correlations between the CFIS and various other criterion measures, including school grades, general intelligence test scores, and specific types of aptitude measures. The researchers have been careful to obtain correlations between the CFIS and the various criterion measures with both caucasian majority group students, and a diverse group of noncaucasian minority students. The following table shows the correlations between the CFIS and the different criterion variables. Remember that a correlation of 1.00 is a *perfect* correlation, while a correlation of 0 indicates that *no* correlation exists.

| Correlations Between the "CFIS" and Criterion Measures | | |
|---|---|---|
| Criterion Variable | Correlation for Majority Students | Correlation for Minority Students |
| School Grades | .38 | .40 |
| General Intelligence | .52 | .50 |
| Verbal Aptitude | .32 | .36 |
| Mathematical Aptitude | .44 | .41 |
| Spatial Relations Aptitude | .84 | .85 |
| Visual Design Aptitude | .79 | .76 |
| Vocabulary Aptitude | .31 | .29 |
| Behavioral Problems | .11 | .09 |

Carefully look over the information provided in the table, then answer the following questions:

1. Does the CFIS predict school grades differently for majority and minority students?

2. Based on the information in the table, what does the CFIS appear to be measuring?

3. Do you think there is evidence that the CFIS effectively removes the influence of extraneous cultural variables and produces a pure measure of aptitude?

You may check your answers against the answers provided at the end of the chapter.

## ANSWERS TO GUIDED STUDY EXERCISES

### Definition Exercises

*Aptitude Tests*: Tests designed to measure specific areas of aptitude, with the goal of being able to predict future performance in that area.

*Culture-Free Tests*: Aptitude tests that are designed to eliminate the effects of extraneous cultural influences so that resulting scores are a pure measure of aptitude.

*Predictive Validity*: The ability of a test to predict future performance by the examinee in a given area.

*Intelligence Tests*: Special kinds of aptitude tests that are designed to measure overall or general cognitive ability.

*Multiple Aptitude Tests*: Special kinds of aptitude tests that are designed to differentially measure specific cognitive abilities.

*School Readiness*: The attainment of prerequisite skills, attitudes, and motivation that enable the learner to profit maximally from school instruction.

*Creativity*: Creativity involves ingenuity, originality, and inventiveness--it is related to intelligence but is difficult to adequately measure.

**Multiple-Choice Questions**

1. C
2. B
3. A
4. D
5. A
6. D
7. A
8. B

**Short-Answer Questions: Key Points**

1. Aptitude, which is natural or acquired ability, is related to and can predict academic achievement to a point, but it is not the total cause of achievement. Other factors, such as motivation and personal problems, can also have an impact on achievement.

2. Efforts to remove the influence of cultural variables from aptitude tests have not been very successful. Even the most extensively researched "culture-free" tests have generated little evidence of advantage or superiority to traditional intelligence tests.

3. Aptitude is a term used to describe specific types of ability. Intelligence is a form of aptitude that refers to general intellectual capacity.

4. The GRE has been shown to predict performance in graduate school to a greater extent than does undergraduate grade point average.

5. Creativity is very difficult to measure because it is so hard to define operationally.

# APPLICATION ACTIVITY ANSWERS

## Deciding What a Test Really Measures

1. Does the CFIS predict school grades differently for majority and minority students?
   *No. The correlations between CFIS scores and school grades for both majority and minority students are essentially the same and are fairly modest.*

2. Based on the information in the table, what does the CFIS appear to be measuring?
   *The CFIS could be best described as a measure of nonverbal intelligence, or more specifically, a test of spatial-visual aptitude. Its low correlations with various types of verbal aptitude scores and high correlations with spatial relations and visual design aptitude suggest that this is so.*

3. Do you think there is evidence that the CFIS effectively removes the influence of extraneous cultural variables and produces a pure measure of aptitude?
   *There is really no evidence in the data presented here that suggests the CFIS is a pure measure of aptitude, with no cultural influences. The data reported here suggest that the CFIS is a test of certain nonverbal abilities that correlate only modestly with academic achievement and general intelligence.*

# CHAPTER 20
# BEING SENSITIVE TO YOUR STUDENTS' PERSONAL PROBLEMS: MEASURES OF PERSONAL AND SOCIAL ADJUSTMENT

## CHAPTER SUMMARY

Classroom teachers must deal with the issues of interpersonal relations, classroom social climate, and personal adjustment in the classroom on a daily basis. This chapter will provide you with a basic understanding of why these variables are important and how you can measure them. The chapter begins with a brief overview of how personal and social adjustment interrelate in the teaching-learning process and then outlines six different areas of social-emotional assessment about which you, as a potential classroom teacher, need to be aware: sociometrics, classroom climate, self-concept, locus of control, test anxiety, and personality.

Some of the assessment areas outlined in this chapter are more applicable to classroom teachers than others. Sociometric choice measures, in particular, should be understood by the classroom teacher. These measures are relatively easy to administer and score and can provide a wealth of information about patterns of friendship and social status in your classroom. Classroom climate measures can also be very useful in analyzing the social climate in your classroom and in developing plans for improving it.

Some other forms of social-emotional assessment, like locus of control and personality measures, are more likely to be used by psychologists than teachers, but you will benefit from understanding how they work. Over the course of a career, it is likely that every classroom teacher will work with at least a few students with serious emotional and behavioral problems who will be referred to psychologists for a more complex assessment of their personal and social adjustment. When a situation like this occurs, it will be helpful for you to have a basic understanding of these more complex social and emotional measures.

## CHAPTER OUTLINE

Social Acceptance and Adjustment in the Classroom
Uses for Social and Personal and Adjustment Measures
Sociometric Choice Measures
    Nomination Measures
    Guess Who Measures
Classroom Climate Measures
Self-Concept Measures
Locus of Control Measures
Test Anxiety Measures
Personality Inventories

# GUIDED STUDY EXERCISES

## Definition Exercises

Locate the following terms in your text and briefly define each in the context of educational measurement and evaluation.

Personal Adjustment:

Social Adjustment:

Sociometric Measures:

Sociogram:

Clique:

Environmental Press:

Classroom Climate:

Self-Concept (or Self-Esteem):

Internal Locus of Control:

External Locus of Control:


Test Anxiety:


Personality Trait:


**Multiple-Choice Questions**

Circle the letter of the item that best answers each question.

1. Where would a teacher find descriptions and reviews of measures of personal and social adjustment for the classroom?

    a. The *Mental Measurements Yearbook*
    b. ETS *Test Collection Bibliographies*
    c. Professional journals
    d. All of the above

2. Which of the following is the most commonly used sociometric method?

    a. The nominating technique
    b. The "guess who" technique
    c. The roster rating technique
    d. Picture sociometric techniques

3. Based on commonly used criteria for classifying sociometric choice measures, a student who is never chosen would be referred to as

    a. a rejectee.
    b. an isolate.
    c. a neglectee.
    d. a cipher

4. A disadvantage of using structured behavioral observation to measure personal and social adjustment is that it

   a. is very costly.
   b. is generally ineffective.
   c. has low reliability or consistency.
   d. has not been validated for such purposes.

5. The set of perceptions we have about ourselves combine to form our

   a. ego.
   b. self-awareness.
   c. self-concept.
   d. personality.

6. The relationship between self-concept and academic achievement is best characterized as

   a. weak.
   b. significant.
   c. negative.
   d. nonexistent.

7. How should self-concept measures be regarded as indicators of a student's level of self-concept?

   a. As tentative indicators
   b. As conclusive indicators
   c. As poor indicators
   d. As substandard indicators

8. Which of the following is a main component of test anxiety?

   a. Low cognitive ability
   b. Emotionality
   c. External locus of control
   d. Cautiousness

## Short-Answer Questions

Provide a brief answer to each of the following questions in the space provided.

1. Briefly compare and contrast the peer nomination and guess who sociometric procedures.

2. How would you address the concerns of parents or administrators who do not want you to use sociometric measurement techniques because they believe it will lead to further rejection and isolation of certain children?

3. What type of adjustment do people make when their behavior is not in harmony with their self-concept?

4. Why should professionals use caution in interpreting self-concept measures?

5. Compare and contrast internal locus of control and external locus of control.

## APPLICATION ACTIVITIES

### Plotting a Sociogram

It is early October, and Ms. Cantwell has been working with this year's group of third graders for over a month. She is very concerned about how the boys in her class are getting along with each other. This year's group has been doing more than the usual amount of name-calling and fighting, and they are not mixing with each other very well. Ms. Cantwell has decided to get some objective information on the boys' social relationship patterns and does this by administering two different sociometric procedures. The first procedure is a positive nomination technique, where she asks the boys to "name two boys that you would like to be friends with." The second procedure is a negative nomination technique to show patterns of social rejection. Ms. Cantwell asked the boys to "name another boy that you would not want to play with." Their responses to the two procedures are as follows:

| Student | Positive Nominee 1 | Positive Nominee 2 | Negative Nominee |
|---|---|---|---|
| Allen | Cory | Steven | Austin |
| Justin | Tru | Juan | Lupe |
| Cory | Juan | Steven | Daniel |
| John | Steven | Cory | Ben |
| Austin | Steven | Cory | Lupe |
| Calvin | Ben | Jason | Lupe |
| Daniel | Cory | Juan | Jason |
| Tru | Glen | John | Austin |
| Steven | Juan | Cory | John |
| Ben | Jason | Calvin | Austin |
| Lupe | Steven | John | Tru |
| Jason | Calvin | Ben | Lupe |
| Glen | Cory | Juan | Austin |
| Juan | Steven | Daniel | Austin |
| Richard | Cory | John | Lupe |

Using the information and examples from the "Nomination Measures" section of Chapter 20, plot a sociogram for the positive nominations procedure. Then, using both the positive and negative nomination information, classify the data according to the seven frequently used classifications (Star, Isolate, etc.). Since the subjects are all boys, there will not be any Cross-Sex Choices, but see how many of the other six categories fit. An example classification breakdown and a positive nomination sociogram are presented following the answers section.

## ANSWERS TO GUIDED STUDY EXERCISES

### Definition Exercises

*Personal Adjustment*: The combination of an individual's personality traits and self-concept, and how it affects his or her personal progress and social-emotional health.

*Social Adjustment*: An individual's pattern of friendships, social acceptance, social rejection, or isolation, and how it affects his or her overall level of progress.

*Sociometric Measures*: Tests that show patterns of social acceptance or rejection within a peer group.

*Sociogram*: A display of sociometric choice patterns of students as plotted by the teacher.

*Clique*: A small group of students who are friends with each other and have few if any friendship choices outside of the group.

*Environmental Press*: The composite of pressures in a person's surroundings.

*Classroom Climate*: The pressure a student perceives within a classroom.

*Self-Concept*: The multifaceted set of perceptions each person holds about himself or herself.

*Internal Locus of Control*: A pattern where an individual believes that his or her own actions largely determine rewards and punishments.

*External Locus of Control*: A belief pattern where an individual attributes his or her successes and failures to external forces such as luck, fate, or other persons.

*Test Anxiety*: A condition when a student knows the course material but cannot demonstrate it on a test due to problems of emotionality and worry.

*Personality Trait*: Patterns of long-term and enduring behavioral and emotional characteristics in individuals.

## Multiple-Choice Questions

1. D
2. A
3. B
4. A
5. C
6. B
7. A
8. B

## Short-Answer Questions: Key Points

1. Peer nomination techniques usually ask students to name other classmates that they prefer on some criterion, such as who they prefer to play with. The guess who technique is also a sociometric choice measure, but varies from the traditional nomination technique by having students select the names of their peers who fit particular descriptions (e.g., "Guess who is in trouble?").

2. Perhaps the best way to address the concerns of those who believe sociometric measurement will lead to further rejection and isolation of children is to emphasize that there is no empirical or research basis for such a concern. It has not been demonstrated through research that such an effect exists in sociometric assessment.

3. When a person's behavior and self-concept are inconsistent, either behavior must change or self-concept must change. If this does not happen, a high degree of subjective stress may result.

4. One of the major concerns in using self-concept measures is that they are difficult to validate, and it is unclear what particular score patterns might indicate.

5. Individuals with an internal locus of control view what happens to them as being a result of what they do. Individuals with an external locus of control believe that such things as luck and fate are the cause of their problems and achievements.

# APPLICATION ACTIVITY ANSWERS

## Classification Categories

*Stars*: Cory, Steven, and Juan. Each received at least five positive nominations.

*Isolates*: Allen, Glen, Richard, Justin. None of these students received any positive or negative nominations.

*Neglectees*: Tru and Daniel. Each received only one positive nomination.

*Rejectees*: Lupe and Austin. Each received five negative nominations but no positive nominations.

*Mutual Choice*: The following pairs of students selected each other in the positive nominations: Jason-Calvin, Jason-Ben, Calvin-Ben, Daniel-Juan, Steven-Juan, and Steven-Cory.

*Clique*: Jason, Calvin, and Ben selected each other and no one else for positive nominations.

## An Example of a Sociogram for the Positive Nomination Data

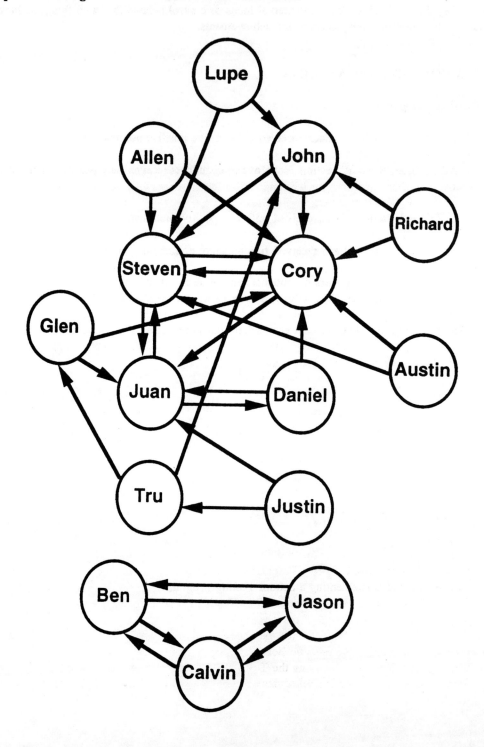

# CHAPTER 21
# SETTING UP A SCHOOL TESTING PROGRAM

## CHAPTER SUMMARY

Most of the information and discussion on tests from your text have focused on how to select, use, and interpret tests that have applications for specific classroom purposes. This focus makes sense because these are the purposes in which teachers are most interested. There are, however, other reasons for administering tests to your students. Chapter 21 in your text outlines and details how to make good use of tests within the framework of an entire school testing program. Such programs will ideally help in achieving both classroom purposes for teachers and in larger schoolwide purposes that are of interest to administrators, school board members, and the like.

Chapter 21 begins with a discussion of why comprehensive testing programs are important and what educational purposes these programs will serve. Next, 14 different "ingredients" in a good testing program are outlined and discussed, based on what a well-conceived testing program should be able to accomplish. The next section of Chapter 21 outlines and details eight specific steps to use in developing and implementing a school testing program. These eight steps range from determining what purposes your testing program is to serve, to developing procedures for periodic review and evaluation of the testing program. The chapter ends with an extensive discussion of the use of microcomputers in school testing programs and gives detailed guidelines to follow in selecting both computer hardware and software applications that can be used for school testing programs.

In times past, most classroom teachers rarely needed to be concerned about setting up schoolwide or districtwide testing programs, but times have changed. With the increased emphasis on site-based management and collaborative decision making between teachers and administrators that has been seen within the past decade, teachers now need to understand the basic workings of school testing programs. This chapter will provide you with the information and skills you need to develop such an understanding.

## CHAPTER OUTLINE

The Importance of a Comprehensive Testing Program
Important Ingredients in a Good Testing Program
Steps in Developing and Implementing a School Testing Program
        Step 1: Determining What Purposes Your Testing Program Is to Serve
        Step 2: Choosing the Level of the Testing Program
        Step 3: Developing a Structure for the Overall Testing Program
        Step 4: Determining the Specific Behaviors and Content to Be Measured
        Step 5: Choosing or Developing the Tests to Be Used
        Step 6: Developing Testing Schedules and Guidelines for Test Administration

Step 7: Developing Procedures for Test Scoring, Analysis, and Reporting
Step 8: Developing Procedures for Periodic Review and Evaluation of the Testing Program
Using Microcomputers in Your Testing Program
The Problem of "Computer Anxiety"
The Use of Microcomputers in School Testing Programs
Deciding What Microcomputer System Fits Your Needs

## GUIDED STUDY EXERCISES

### Definition Exercises

Since Chapter 21 does not introduce any new technical terms or concepts, there are no definition exercises for this workbook chapter.

### Multiple-Choice Questions

Circle the letter of the item that best answers each question.

1. Historically, which group of educators has been least responsible for planning schoolwide or districtwide testing programs?

    a. Administrators
    b. Teachers
    c. School psychologists
    d. Curriculum or testing specialists

2. Which of the following would be considered an administrative purpose for a comprehensive testing program?

    a. Grading of individual student performance
    b. Career counseling
    c. Identifying and solving educational problems
    d. Evaluating curricula or programs

3. The practice of conducting districtwide achievement testing programs in both the spring and fall

    a. is seldom needed.
    b. is sound educational practice.
    c. is seldom done in actual practice.
    d. does not help to determine how much "forgetting" occurs over the summer.

4. Well-designed testing programs that provide simultaneous, parallel reports are used to provide

   a. individual and aggregated test data.
   b. student portfolio information.
   c. information on both student and teacher behavior.
   d. ordinal and nominal information.

5. For districtwide testing programs to be useful,

   a. they must evaluate teacher as well as student outcomes.
   b. the core curriculum must be consistent across schools.
   c. they must be strictly norm referenced.
   d. the selection of tests should be done by classroom teachers rather than specialists.

6. Compared with elementary teachers, secondary teachers test

   a. about the same amount.
   b. considerably less.
   c. about four times as much.
   d. about twice as much.

7. Minimum competency tests can present a problem because educators

   a. tend to dislike using them.
   b. are asked to set standards for groups rather than individuals.
   c. often set unreasonably high passing standards.
   d. are usually forced to develop these tests based on legal rather than educational criteria.

8. Which of the following is the biggest problem with use of microcomputers in schools?

   a. Most schools still do not have access to computers.
   b. Students tend to be anxious and intimidated about using computers.
   c. Most educational software programs are of very poor quality.
   d. Computers are not effectively used for instructional purposes.

## Short-Answer Questions

Provide a brief answer for each of the following questions in the space provided.

1. What are some reasons that teacher-made tests are seldom integrated effectively into overall district testing programs?

2. Outline how a comprehensive testing program can be important for instructional purposes.

3. How might an individual teacher or a school district go about improving the linkage between testing and instruction?

4. At what grade levels is the administration of academic aptitude and general aptitude tests recommended?

5. What are the main applications of microcomputers in school testing programs?

## APPLICATION ACTIVITIES

### What's Wrong with This Test?

To help you integrate your understanding of how to use school testing programs, read the following scenario about how a hypothetical school district designed and implemented a testing program. Then, make a list of *what things they did right* and *what things they did wrong.* Use the 14 important ingredients in a good testing program from your text as a guide for this activity. Check your responses with the answers that are provided at the end of this chapter. As you will see, the use of school testing programs is hardly ever an "all or nothing" proposition--usually, they have both positive and negative aspects.

The school board of the Maple Valley School District charged a committee consisting of a curriculum specialist and a group of three teachers with the task of selecting and implementing a districtwide academic assessment program. They were specifically charged with the responsibility of implementing a system that will measure student progress *in all areas important to student academic progress*. After reviewing several test programs, the committee decided to adopt the Test of Basic Academic Skills (TBAS), a test that focuses on basic reading and math skills at all grade levels. The committee adopted the TBAS after consultation and input from many parents, teachers, and administrators. The TBAS was selected primarily for its ease of use, cost, and strong psychometric properties, but not for its relationship to local curriculum goals (which is fairly weak). One of the advantages of the TBAS is that it provides a high degree of diagnostic and prescriptive information, both individually and at different grade levels. The curriculum specialist and superintendent agreed that testing with the TBAS would occur in September and again in May, "so that we can provide the board and the community with a lot of information about how our students are doing." The TBAS was administered at the agreed-on times (September and May), but scores were not made available to teachers and parents until the end of November for the September tests, and the following September for the May tests. The superintendent and school board were very pleased with the overall implementation and outcome of the TBAS testing and decided to use it the following year. One school board member was so impressed with the information the TBAS provided that she convinced the school board that "we ought to subtly convince the building principals to look at and use TBAS scores of entire classrooms when teachers are being given their annual performance evaluations."

## ANSWERS TO GUIDED STUDY EXERCISES

### Multiple-Choice Questions

1. B
2. D
3. A
4. A
5. B
6. D
7. B
8. D

### Short-Answer Questions: Key Points

1. Teacher-made tests are seldom integrated effectively into the overall district testing program because teachers typically have limited input into the decision to purchase districtwide tests.

2. The most important aspect of making a comprehensive testing program useful for instructional purposes is to select tests with a high match to the curriculum.

3. The linkage between testing and instruction can be improved through (a) building new criterion-referenced tests, (b) modifying existing tests by removing items not relevant to the curriculum, (c) constructing taxonomies of tests and curricula in order to develop selection guidelines, and (d) measuring the degree of overlap in existing tests and curricula.

4. It has been recommended that academic aptitude tests be administered during grade 3 or grade 4. The recommendation for general aptitude tests is that they be administered during grade 10 or grade 11.

5. The most common use of computers in testing involves adapting traditional paper-and-pencil tests to a computerized format. Unfortunately, this is one of the least effective uses of computers in testing.

## APPLICATION ACTIVITY ANSWERS

**What's Wrong with This Test?**

*Things They Did Right*

1. Developed the program cooperatively with teachers and other educators

2. Selected a test that supports instructional functions for groups of students

3. Selected a test that permits diagnosis of individual student performance

4. Selected a test system that enhanced the interpretability of test results across grade levels

5. Met public needs for school accountability by making scores publicly available

*Things They Did Wrong*

1. Selected a test system that did not provide for measurement of the most important outcomes (it was limited to only reading and math)

2. Did not link testing and instruction

3. Engaged in undue repetition or redundancy by testing twice a year

4. Did not provide timely and appropriate test information

5. By deciding to use TBAS scores for teacher evaluation, used the test system for other than legitimate, agreed-upon purposes

# CHAPTER 22
# SETTING UP A SCHOOL EVALUATION PROGRAM

## CHAPTER SUMMARY

The last chapter in your textbook is devoted to a topic that is a slight departure from the previous chapters. While most of the content of the text has been focused on evaluation of individual student performance, Chapter 22 expands the meaning of evaluation to include the purposes and methods of evaluating school programs. In this sense, the broad definition of evaluation is enlarged to include the broader scrutiny of particular educational curricula, programs, projects, and methods.

Chapter 22 begins with a discussion of how evaluation is used for programs rather than for individual student performance. Next, the chapter presents the basic concepts and methods of evaluation and how it is actually used to improve school programs. The third section of the chapter compares and contrasts six alternative approaches to evaluating school programs. Following this comparative overview, the text presents detailed guidelines for setting up a school evaluation program. The chapter ends with some suggestions for obtaining useful evaluation help from sources outside of your school or agency.

Chapter 22 points out that many teachers may have participated in school evaluation programs and not realized it. It is likely that any educator who spends a fair amount of time in the profession will be involved at some level in school evaluation activities. Chapter 22 and the exercises and activities in this workbook will help you become more familiar with the concept of program evaluation and will provide you with some solid general guidelines to use should you ever become responsible for conducting formal evaluation activities.

## CHAPTER OUTLINE

Evaluation of Programs, Not of Individual Student Performance
Using Evaluation to Improve School Programs
    Informal and Formal Evaluation
    Two Basic Distinctions in Evaluation
    Measurement as a Key Evaluation Tool
Guidelines for Setting Up a School Evaluation Program
    Deciding When to Conduct an Evaluation
    Deciding What (Precisely) to Evaluate
    Deciding Who the Evaluation is For
    Deciding Who Should Conduct the Evaluation
    Deciding What Questions the Evaluation Should Address
    Planning the Collection and Analysis of Information
    Conducting and Reporting an Evaluation
    Dealing with Ethical Issues in Evaluation Studies
Suggestions for Obtaining Useful Evaluation Help from Outside Sources

# GUIDED STUDY EXERCISES

## Definition Exercises

Locate the following terms in your text and briefly define each in the context of educational measurement and evaluation.

Evaluation:

Formative Evaluation:

Summative Evaluation:

Internal Evaluation:

External Evaluation:

Objectives-oriented Evaluation:

Management-oriented Evaluation:

Consumer-oriented Evaluation:

Expertise-oriented Evaluation:

Adversary-oriented Evaluation:

Participant-oriented Evaluation:

Cyclical Evaluation Systems:

Stakeholders:

**Multiple-Choice Questions**

Circle the letter of the item that best answers each question.

1. Informal evaluation methods

    a. are generally preferable since they are less costly.
    b. have limited value in decision making because they are based on highly subjective perceptions.
    c. are systematic efforts to collect and weigh accurate information about alternatives.
    d. tend to be more useful for decision making than formal evaluation methods.

2. Which of the following types of evaluation is best used to make "go/no go" decisions?

    a. Summative
    b. Internal
    c. Formative
    d. External

3. An evaluation of a school district's primary-grade reading program, conducted by the district curriculum specialist for the purpose of identifying areas of the curriculum that could be improved, is an example of which of the following types of evaluation?

    a. External-summative
    b. Internal-summative
    c. External-formative
    d. Internal-formative

4. Which of the following types of evaluation rarely happens in education?

   a. Internal-formative
   b. External-formative
   c. Internal-summative
   d. External-summative

5. The Heartland Community College recently went through an evaluation process for renewal of their accreditation. This process involved outside professionals coming in to the school to make judgments in different areas. Which evaluation approach was used?

   a. Objectives-oriented
   b. Consumer-oriented
   c. Management-oriented
   d. Expertise-oriented

6. The Middleton School District recently conducted an evaluation of their alternative high school, which is a controversial issue with the school board. They selected two individuals to assist with the evaluation--one who was known to have positive views on alternative schools, and one who was known to have negative views. Which evaluation approach was used?

   a. Consumer-oriented
   b. Adversary-oriented
   c. Objectives-oriented
   d. Expertise-oriented

7. Which of the following is *not* a potential problem with using a cyclical evaluation schedule?

   a. If the evaluations are on a set cycle, information may be "contrived" to show good outcomes.
   b. Some districts have practical constraints that prevent them from establishing a regular schedule.
   c. Issues can arise that require changes in original schedules.
   d. After a long-range evaluation cycle is established, it is possible that the cycle may not be effective.

8. Which of the following statements is true regarding certification/licensure systems for educational program evaluation specialists?

   a. Most states require a doctoral degree in educational evaluation for certification.
   b. There is no widespread system for certification or licensure of program evaluators.
   c. Twenty-eight of the 50 U.S. states have certification/licensure systems for program evaluators.
   d. The majority of states require a master's degree in educational evaluation for certification.

## Short-Answer Questions

Provide a brief answer for each of the following questions in the space provided.

1. Compare and contrast summative evaluation and formative evaluation. What are the purposes, advantages, and uses of each of these two types of evaluation?

2. When should external evaluation be the preferred method rather than internal evaluation?

3. Describe a situation where it would be desirable to obtain outside help in planning or conducting an evaluation.

4. Which of the six common evaluation approaches would be most useful for a formative evaluation of an educational curriculum within one school? Why?

5. How does one go about deciding who the "stakeholders" are for a given evaluation?

## APPLICATION ACTIVITIES

### What Kind of Evaluation Is This?

Chapter 22 in your text introduces you to the evaluation dimensions of *formative* versus *summative* and *internal* versus *external*. Completing this activity will help you to better understand the distinctions along these dimensions and to identify specific types of combinations along the following breakdown: (a) Internal-formative, (b) external-formative, (c) internal-summative, and (d) external-summative.

First review the section of Chapter 22 that addresses the distinctions along these evaluation domains. Then, classify each of the four examples in the table that follows.

| APPLICATION ACTIVITY: FOUR EXAMPLES OF SPECIFIC EVALUATION TYPES (answers are provided at the end of the chapter) ||
|---|---|
| Example | Evaluation Classification |
| An educational corporation has recently developed and published a new high school algebra curriculum. They assign one of their project staff to conduct an evaluation of the curriculum by comparing it with three widely used algebra curricula. The results of the evaluation are released to educators as a selling point for the new curriculum. | |
| School district A is concerned with how computers are being used in the elementary schools. They are interested in making some changes in the way they operate the computer program in order to better facilitate learning objectives. They invite an individual from a nearby community who is widely regarded as an expert on computer applications and learning to evaluate the way the computers are being used and to make recommendations for change. | |
| School district B is using two different elementary reading curriculum programs and has decided to make a determination of which program is most effective and to use that one only. They ask Drs. J and H, two elementary education professors at the state university, to conduct an evaluation of the two curricula and to make a recommendation as to the merits and weaknesses of each. | |
| A middle school has been utilizing an advisory program for the past two years, assigning each teacher a group of students during the first 20 minutes of the school day in order to increase positive role modeling and better student attitudes toward school. There are some "bugs" in the program, and the principal assigns the school counselor to conduct an evaluation of the program in order to make recommendations for improvements to the advisory system. | |

# ANSWERS TO GUIDED STUDY EXERCISES

## Definition Exercises

*Evaluation*: The determination of worth, value, or quality. As the term is used in Chapter 22, it specifically relates to institutional or programmatic performance (curricula, programs, projects, methods), as opposed to evaluation of individual performance.

*Formative Evaluation*: Evaluation conducted during the planning and operation of a school program that provides those involved with evaluative information that can be used in improving the program.

*Summative Evaluation*: Evaluation that occurs after a curriculum or program is ready for regular use; it provides potential consumers with evidence about the program's worth.

*Internal Evaluation*: Evaluation that is conducted by someone within the school or system where the program is based.

*External Evaluation*: Evaluation that is conducted by someone from outside the school or system where the program is based.

*Objectives-oriented Evaluation*: Evaluation that focuses upon specifying goals and objectives and determining the extent to which they have been obtained; it is essentially a comparison between performance measures and objectives.

*Management-oriented Evaluation*: Evaluation that is specifically designed to serve the needs of decision makers; the central concern is identifying the informational needs of decision makers involved within the system and collecting sufficient information about the relative advantages and disadvantages of each alternative.

*Consumer-oriented Evaluation*: Evaluation where the central issue is developing evaluative information on educational "products" for use by educational consumers who are choosing among competing methods, curricula, instructional products, and so forth.

*Expertise-oriented Evaluation*: Evaluation that depends primarily upon professional expertise to judge an educational institution, program, product, or activity.

*Adversary-oriented Evaluation*: Evaluation that incorporates planned opposition (pro and con) between different evaluators, thus attempting to balance the evaluation and ensure fairness by incorporating both positive and negative viewpoints.

*Participant-oriented Evaluation*: Evaluation where the involvement of participants is essential in determining the values, criteria, needs, and data for the evaluation.

*Cyclical Evaluation Systems*: An evaluation "calendar" where evaluation plans and periodic reviews are scheduled in a staggered fashion so that some component of the curricula or program is evaluated every year, and at reasonable intervals.

*Stakeholders*: Persons who have a legitimate interest in the outcomes of an evaluation. In education, examples of stakeholders might include policy makers, administrators, practitioners, students, teachers, and community groups.

## Multiple-Choice Questions

1. B
2. A
3. D
4. B
5. D
6. B
7. A
8. B

## Short-Answer Questions: Key Points

1. Formative evaluation is conducted during the planning and operation of a school program and provides those involved with evaluative information that can be used in improving the program. Summative evaluation, on the other hand, occurs after a curriculum or program is ready for regular use; it provides potential consumers with evidence about the program's worth.

2. External evaluation should be used over internal evaluation when (a) having the evaluation conducted internally would involve a conflict of interest, or (b) no one is available internally with the necessary expertise to conduct the evaluation.

3. Obtaining outside help in conducting an evaluation would be desirable in a case where an outside specialist can provide valuable consultation on some of the technical issues in conducting the evaluation. An example would be if the evaluation were being conducted on an area where there was limited expertise within the system.

4. Of the six common evaluation approaches, the participant-oriented evaluation would probably be the most useful method for conducting a formative evaluation of a curriculum within one school. This method allows for participation of those who will be affected by the evaluation and provides opportunity for input on important issues such as "how can we best adapt this curriculum?"

5. Although there is no specific formula or method for correctly identifying who the "stakeholders" in an evaluation are, common sense would dictate that they could be identified by simply deciding what the possible outcomes of the evaluation might be and then determining which groups (policy makers, administrators, developers, practitioners, primary or secondary consumers) are likely to be affected by the outcomes.

# APPLICATION ACTIVITY ANSWERS

| FOUR EXAMPLES OF SPECIFIC EVALUATION TYPES: ANSWERS ||
|---|---|
| Example | Evaluation Classification |
| An educational corporation has recently developed and published a new high school algebra curriculum. They assign one of their project staff to conduct an evaluation of the curriculum by comparing it with three widely used algebra curricula. The results of the evaluation are released to educators as a selling point for the new curriculum. | Internal-Summative |
| School district A is concerned with how computers are being used in the elementary schools. They are interested in making some changes in the way they operate the computer program in order to better facilitate learning objectives. They invite an individual from a nearby community who is widely regarded as an expert on computer applications and learning to evaluate the way the computers are being used and to make recommendations for change. | External-Formative |
| School district B is using two different elementary reading curriculum programs and has decided to make a determination of which program is most effective and to use that one only. They ask Drs. J and H, two elementary education professors at the state university, to conduct an evaluation of the two curricula, and to make a recommendation as to the merits and weaknesses of each. | External-Summative |
| A middle school has been utilizing an advisory program for the past two years, assigning each teacher a group of students during the first 20 minutes of the school day in order to increase positive role modeling and better student attitudes toward school. There are some "bugs" in the program, and the principal assigns the school counselor to conduct an evaluation of the program in order to make recommendations for improvements to the advisory system. | Internal-Formative |

# NOTES